Managing Complex LNG Projects throughout Their Life-Cycle

Managing Complex LNG Projects throughout Their Life-Cycle serves as an invaluable resource for both aspiring and experienced professionals in the field of LNG project management. It offers practical insights and strategies essential for navigating the complexities inherent in mega LNG projects. Readers will gain a comprehensive understanding of best practices and critical success factors, equipping them to effectively manage and deliver successful LNG initiatives. The management of complex LNG projects stands at the intersection of innovation, investment, and global energy demands, yet, amidst their developments, these ventures pose unprecedented challenges.

By exploring real-world case studies and expert analyses, this book not only enhances theoretical knowledge but also provides actionable tools for achieving project goals in a challenging and dynamic industry. From initial planning and risk assessment to execution and monitoring, the book offers practical strategies and tools to help mitigate risks, control costs, and ensure timely delivery. Ultimately, the aim is to empower project teams with the knowledge and resources they need to deliver successful outcomes in the challenging landscape of mega LNG projects.

By carefully understanding the intricacies of each phase, from initial planning and feasibility studies to construction, commissioning, and operation, this book can offer targeted strategies and best practices for effective project management and is essential reading for oil and gas professionals, engineers, and project managers worldwide.

Dr Noel J. Mitchell has over 40 years' experience in the onshore/offshore oil and gas industry, working in senior positions developing oil and gas projects for many organisations, including Shell, BP, Caltex, Fluor, Woodside, Chevron, Total, Turner Townsend, and others.

Managing Complex LNG Projects throughout Their Life-Cycle

Dr Noel J. Mitchell

Routledge
Taylor & Francis Group
LONDON AND NEW YORK

Designed cover image: Dr Noel J. Mitchell

First published 2026
by Routledge
4 Park Square, Milton Park, Abingdon, Oxon OX14 4RN

and by Routledge
605 Third Avenue, New York, NY 10158

Routledge is an imprint of the Taylor & Francis Group, an informa business

British Library Cataloguing-in-Publication Data
A catalogue record for this book is available from the British Library

ISBN: 978-1-041-16689-4 (hbk)
ISBN: 978-1-041-17169-0 (pbk)
ISBN: 978-1-003-68833-4 (ebk)

DOI: 10.1201/9781003688334

Typeset in Times New Roman
by Apex CoVantage, LLC

I would like to give my special thanks to my wife, Margaret, for her great support throughout this long journey to develop this book. This includes her patience and her assistance in scrutinising this book. I could not have achieved this without her support. Thank you.

Contents

Figures

Tables

Appendixes

Abbreviations

BM:	Benchmarking
BOD:	Basis of Design
BSCs:	Balanced Scorecards
BWI:	Barrow Island
CAPEX:	Capital Expenditure
CBS:	Cost Breakdown Structure
CEO:	Chief Executive Officer
CFA:	Confirmatory Factor Analysis
CIs:	Core Indicators
CM:	Construction Management
COW:	Carry-Over-Work
CSFs:	Critical Success Factors
CSG:	Coal Seam Gas
DSP:	Decision Support Package
EMS:	Engineering Management Services
EPCM:	Engineering, Procurement, Construction, Management
EVMS:	Earned Value Management System
FEED:	Front End Engineering Design
FEL:	Front End Loading
FLNG:	Floating LNG
FID:	Financial Investment Decision
IM:	Interface Management
IPA:	Independent Project Analysis
IPMT:	Integrated Project Management Team
JVPs:	Joint Venture Partners
KPIs:	Key Performance Indicators
LNG:	Liquefied Natural Gas
Mid-Stream:	Subsea Installations (xmas trees, pipelines, etc.)
MOF:	Material Offloading Facilities
NG:	Natural Gas
NPV:	Net Present Value
NWS:	Northwest Shelf
OG:	Oil and Gas

OPEX: Operational Expenditure
PARs: Pre-Assembled Racks
PAUs: Pre-Assembled Units
PBB: Phased-Based Benchmarking
PC: Project Charter
PCA: Principal Component Analysis
PEP: Project Execution Plan
PF: Performance Forum
PMBOK: Project Management Book of Knowledge
PMP: Project Management Plan
PMS: Project Management System
RFSU: Ready for Start Up
SoW: Scope of Works
SWOT: Strengths, Weaknesses, Opportunities, and Threats
TQM: Total Quality Management
V*f*M: Value *for* Many
WA: Western Australia
WBS: Work Breakdown Structure

About the Author

With over 40 years of extensive experience in the onshore and offshore oil and gas industry, the author has developed a deep understanding of the sector's complexities and challenges. Throughout his career, he has held leadership roles in project management, driving innovation and efficiency in diverse environments.

Having worked in multiple regions globally, including high-risk offshore locations, with a proven track record of delivering large-scale projects on time and within budget, while consistently maintaining the highest safety and environmental standards. His career has been marked by a commitment to continuous learning, planning, cost control, and adapting to evolving technologies and industry trends. His technical acumen, combined with strategic vision, has contributed to the successful execution of both upstream and downstream projects. Passionate about mentoring the next generation of energy professionals, the author also dedicates time to training and knowledge transfer, ensuring the ongoing development of industry talent in developing multifaceted projects, including mega complex LNG projects.

He has worked in senior positions on offshore oil and gas platforms, onshore facilities, module fabrication yards (South Korea and China), FPSOs refits, FPU fabrication, etc.

He has gained various academic qualifications in project management, including an MSc and a PhD. The PhD thesis explores how organisations can effectively manage the complexities of mega LNG projects throughout their life-cycle to ensure success. Drawing on extensive research and practical insights, the thesis addresses key challenges such as project planning, risk management, stakeholder alignment, and performance optimisation, providing a comprehensive framework for navigating the intricate dynamics of large-scale LNG ventures.

This in-depth exploration of complex project management in the LNG sector inspired the author to share his wealth of knowledge and real-world experience with a broader audience. Motivated by the desire to help others succeed in the industry, the author wrote this book as a practical guide, distilling decades of hands-on expertise into actionable strategies, lessons learnt, and best practices for managing mega LNG projects from inception to completion.

Preface

This book aims to equip professionals and organisations with the tools and insights needed to overcome the unique challenges of LNG project development, delivering value while minimising risks and ensuring sustainable outcomes.

Consequently, it serves as an invaluable resource for both aspiring and experienced professionals in the field of LNG project management. It offers practical insights and strategies essential for navigating the complexities inherent in mega LNG projects. Readers will gain a comprehensive understanding of best practices and critical success factors, equipping them to effectively manage and deliver successful LNG initiatives. By exploring real-world case studies and expert analyses, this book not only enhances theoretical knowledge but also provides actionable tools for achieving project goals in a challenging and dynamic industry.

It focuses on several key themes related to the development of mega LNG projects throughout their life-cycle, such as:

Technological Innovations: Exploring advancements in extraction, transportation, and processing technologies that enhance efficiency and reduce environmental impact.

Risk Management Strategies: Identifying potential risks – financial, operational, environmental – and outlining strategies to mitigate these risks, including regulatory compliance and stakeholder engagement.

Economic Viability: Analysing the financial aspects of LNG projects, including investment costs, market demand, and pricing mechanisms to ensure profitability.

1. **Environmental Impact**: Addressing sustainability concerns and the carbon footprint of LNG projects, including measures for reducing emissions and improving overall environmental performance.
2. **Supply Chain Dynamics**: Examining the complexities of logistics, from sourcing natural gas to delivering LNG to markets, and the importance of robust supply chain management.
3. **Regulatory Frameworks**: Understanding the legal and regulatory landscape that governs LNG projects, including permitting processes and international trade agreements.
4. **Stakeholder Engagement**: Highlighting the importance of collaboration among various stakeholders – governments, investors, local communities – to ensure successful project execution and social acceptance.

5. **Market Trends**: Investigating global market trends, such as shifts in energy consumption, geopolitical factors, and competition from renewable energy sources.

By addressing these themes, the book aims to provide a comprehensive understanding of the challenges and opportunities in developing and managing mega LNG projects.

The management of mega complex LNG projects stands at a complex intersection where innovation, substantial investment, and the global demand for energy converge. These projects often represent some of the largest infrastructure undertakings in the energy sector, and their execution requires a careful balance of technology, planning, financial resources, and geopolitical considerations. Yet, amidst their developments, these ventures pose unprecedented challenges that require thoughtful management at every stage (phase) of the project life-cycle. Issues such as fluctuating market prices, regulatory complexities, and environmental concerns require innovative solutions and adaptive strategies. By addressing challenges head-on with careful planning, technological adoption, and stakeholder collaboration, organisations can ensure the successful delivery and long-term sustainability of such complex projects. Additionally, the transition towards renewable energy sources is a critical factor that adds pressure to traditional industries like LNG. As global energy consumption shifts towards cleaner and more sustainable alternatives, the natural gas sector faces increasing scrutiny and demands for greater accountability in terms of environmental impact, carbon emissions, and long-term viability. This transition requires a careful balancing act between maintaining profitability and embracing sustainability, particularly for mega LNG projects that require massive investments and have long operational life-spans. However, by embracing innovation, prioritising sustainability, and diversifying into renewable energy, LNG organisations can ensure their profitability while aligning with global environmental goals.

Navigating these challenges is crucial for ensuring the future viability of the industry while meeting global energy demands.

Managing mega LNG projects is an immense undertaking, and the outbreak of the COVID-19 pandemic has highlighted the vulnerabilities and uncertainties inherent in such massive undertakings. In the face of unforeseen challenges, this book has provided a framework for effective project management tailored specifically to mega complex LNG projects.

The onset of the COVID-19 pandemic significantly amplified the complexities, underscoring the urgent need for a comprehensive approach to project management. The pandemic introduced unprecedented disruptions to workflows, communication, supply chains, and team dynamics. With remote work becoming the norm and the global economic landscape shifting rapidly, traditional project management methodologies were often ill-suited to address the new challenges. To navigate these complexities, organisations need to embrace a more flexible and responsive approach to project management. The COVID-19 pandemic served as a catalyst for change in project management, pushing organisations to rethink their approaches and adopt more dynamic, technology-driven, and people-focused strategies. The lessons learnt during this time will likely shape the future of project management for years to come, emphasising the need for resilience, adaptability, and a more holistic approach to managing projects in a volatile world.

This book aims to bridge this critical gap in the industry by equipping practitioners with the knowledge and strategies necessary to navigate these challenges effectively. By offering insights into the entire project life-cycle and presenting practical strategies for risk mitigation, budget control, and schedule management, the book equips professionals with the tools they need to tackle the unique challenges of mega LNG projects head-on.

Mega LNG projects often face significant challenges, many of which lead to budget overruns and schedule delays, which can have far-reaching consequences. With significant investments and the involvement of numerous stakeholders, managing risks and uncertainties becomes an essential aspect of project planning and execution.

This book serves as a guide to address these issues by offering insights into the best practices for managing mega complex LNG projects throughout their life-cycle. By examining a case study, analysing past failures and successes, and drawing on industry expertise, it has provided a roadmap to assist project managers to navigate the complexities of mega LNG projects more effectively. From initial planning and risk assessment to execution and monitoring, the book offers practical strategies and tools to help mitigate risks, control costs, and ensure timely delivery. Ultimately, the aim is to empower project teams with the knowledge and resources they need to deliver successful outcomes in the challenging landscape of mega LNG projects.

Examining each phase of the LNG project life-cycle in-depth is crucial for providing valuable insights and guidance to project managers and stakeholders. By carefully understanding the intricacies of each phase, from initial planning and feasibility studies to construction, commissioning, and operation, this book can offer targeted strategies and best practices for effective project management. It delves into the complexities of mega LNG projects, providing insights that are crucial for navigating the multifaceted challenges presented at every stage of the project life-cycle. Through this comprehensive approach, the book aims to equip professionals with the tools needed to enhance project efficiency, mitigate risks, and achieve successful outcomes in an increasingly volatile environment.

For example, during the planning phase, which sets the foundation for a project's success, it provides guidance on conducting thorough risk assessments, stakeholder engagement, and developing comprehensive project plans. During the execute phase, it can offer insights into effective resource allocation, procurement strategies, and change management processes. Furthermore, during the evaluation phase, it can help project teams assess performance, identify lessons learnt, and make informed decisions for future projects.

Ultimately, by leveraging this comprehensive understanding of the LNG project life-cycle, the book can help ensure that projects are managed effectively, risks are mitigated proactively, and objectives are achieved successfully from initiation to operation.

All phases of these diverse LNG projects will be discussed in-depth. However, the main focus will be on the Develop phase to obtain Financial Investment Decision (FID) and the Execute phase, which is known to be the most expensive and longest phase of the project life-cycle.

Focusing in-depth on the Develop and Execute phases of LNG projects aligns with its critical importance and complexity. In the Develop phase, leading up to the FID, is where key decisions are made regarding project viability, cost estimates, regulatory approvals, and securing financing. By thoroughly exploring this phase, the book can provide valuable insights into the strategies and considerations necessary to successfully progress from conceptualisation to FID.

Similarly, the Execute phase is where the bulk of the project's resources and efforts are concentrated. It involves the actual construction, procurement, installation, and commissioning activities necessary to bring the project to fruition. Given the significant cost and duration, closely examining this phase allows for a detailed understanding of the challenges, risks, and best practices involved in managing such large-scale endeavours.

By focusing on these pivotal phases, the book can equip project managers and stakeholders with the knowledge and tools needed to navigate the complexities and mitigate the risks inherent in mega complex LNG projects, ultimately enhancing the likelihood of project success and delivering value to all stakeholders involved.

The Execute phase of mega LNG projects marks the culmination of meticulous planning and preparation, transitioning into active construction and implementation. This phase encompasses a diverse range of expenses essential for project execution, encompassing materials, equipment, procurement, labour costs, transportation, site management, safety measures, and overall construction processes. It explores these critical cost components and strategies for effective expense management to ensure project success and adherence to budgetary constraints. The complexity and economy of scale of mega LNG projects often contribute to the significant costs involved during this phase. Efficient project management, effective risk mitigation strategies, and a well-defined Project Management Plan (PMP) and a Project Execution Plan (PEP) are essential for controlling costs and ensuring successful project delivery.

Keywords: Complexity, Overruns, Benchmarking, Mega, Giga, Schedule, Life-Cycle, COVID-19

Introduction

Managing multifaceted LNG projects is a complex endeavour that involves navigating a web of interconnected elements, which requires a deep understanding of the dynamics between stakeholders, technologies, and global energy markets. The complexities inherent in these projects demand rigorous organisational management strategies. It also delves into these critical aspects, offering insights and frameworks essential for navigating the intricacies of mega complex LNG projects.

The management of multifaceted LNG projects requires a nuanced understanding of various organisational dynamics, industry trends, and regulatory landscapes.

These projects, characterised by their scale, technological sophistication, and global reach, present unique challenges and opportunities for organisations involved. This book examines the complexities of managing mega LNG projects, offering critical insights and discussions essential for understanding the intricate dynamics at play. In the vibrant landscape of LNG project management, organisations face continuous challenges in achieving operational excellence, managing risks, and meeting stakeholder expectations. Benchmarking data serves as a cornerstone for informed decision-making, offering valuable insights into industry standards, best practices, and performance metrics. This book explores the significance of benchmarking in LNG projects, providing practical guidance and a case study to illustrate that it has a transformative impact on organisational effectiveness. However, in some cases, benchmarking data may not be readily available or applicable to the specific context of mega/giga complex LNG projects. In such instances, organisations may rely on tacit knowledge, implicit expertise and insights gained through experience that individuals bring to the table. This tacit knowledge can be invaluable in informing decision-making, problem solving, and navigating complex project challenges.

Exploring the sources and utilisation of benchmarking data, as well as the role of tacit knowledge, can offer a holistic understanding of how organisations effectively manage multifaceted LNG projects. Benchmarking data and tacit knowledge are two invaluable assets that organisations can harness to achieve operational excellence and innovation. This book explores the sources, utilisation, and synergies between benchmarking data and tacit knowledge, offering insights into their pivotal roles in enhancing project management strategies in the LNG sector.

Successful management of LNG projects hinges on the effective utilisation of resources, expertise, and strategic insights to achieve optimal outcomes. By examining both quantitative metrics and qualitative factors, project teams can enhance decision-making, mitigate risks, and foster innovation throughout the project life-cycle. This book explores the synergies between quantitative and qualitative approaches in LNG project management, offering strategies and a case study to illustrate their integrated application.

Mega LNG projects are renowned for their complexity, scale, and global impact. However, they also face significant challenges, including the tendency to exceed budgets and experience schedule delays. Addressing these issues is paramount for ensuring project success, meeting stakeholder expectations, and maintaining competitiveness in the energy market. This book explores a multifaceted approach to mitigating budget overruns and schedule delays in mega LNG projects, emphasising thorough planning, effective risk management, and proactive strategies. By delving into the factors contributing to budget overruns and schedule creeps, this book can help shed light on common pitfalls and challenges faced by organisations developing mega LNG projects. It can also explore best practices and lessons learned from past projects, providing valuable insights for project managers and stakeholders seeking to avoid similar outcomes.

Furthermore, this book explores the pivotal roles of benchmarking data and tacit knowledge in enhancing project performance, minimising risks, and ensuring optimal outcomes. By examining how these elements integrate into project management practices, it provides practical guidance for organisations aiming to mitigate the likelihood of budget and schedule overruns.

Ultimately, by addressing these key concerns head-on and providing actionable recommendations, this book aims to empower project teams with the knowledge and tools they need to deliver successful outcomes in the challenging landscape of mega complex LNG projects.

Each phase of these mega complex LNG projects' life-cycle will be discussed and described in detail, including deliverables required in each phase of the project life-cycle to assist organisations and project managers in developing their projects.

The onset of the COVID-19 pandemic disrupted global supply chains, workforce mobility, and economic stability, posing unprecedented challenges to mega LNG projects worldwide. This book explores the multifaceted impacts of the pandemic on LNG projects and offers strategies for navigating these challenges effectively. By examining lessons learnt and best practices from the pandemic, this book provides insights into adapting project management approaches to ensure resilience and continuity in the face of future uncertainties. Meza et al. (2022; p 2) state, "With all these events underway, in early 2020, an unexpected COVID-19 pandemic broke out in China, forcing all the countries in the world to close borders and reduce the economic activity to contain the contagion."

1 Understanding Mega Complex LNG Projects

1.0 Overview

As organisations face significant challenges in executing mega complex LNG projects within budget and on schedule, the need for effective project management, strategic planning, and risk mitigation has never been more critical. This book serves as a comprehensive resource for professionals involved in the planning and execution of these multifaceted projects. It delves into the intricacies of project management, offering practical strategies and methodologies tailored to the unique challenges presented by large-scale LNG initiatives. This book provides essential guidance for professionals to navigate the complexities of the industry and to achieve successful outcomes. While the complexities of mega LNG projects present formidable challenges, organisations can mitigate risks and improve project outcomes through proactive planning, effective execution strategies, and leveraging technological advancements. Organisations need to understand that with multifaceted projects like mega LNG ventures, they are dealing with interconnected components, diverse stakeholders, and often uncertain variables. This book provides the guidance for professionals, that is necessary to navigate these complexities effectively. Consequently, organisations undertaking these massive ventures require an approach that is agile, adaptive, and capable of navigating these complexities with ease. This book offers strategies and insights designed to meet these needs effectively. Accordingly, managing these multifaceted endeavours necessitates a mindset shift compared to more straightforward projects. Some key mindset changes are:

- understanding and accepting that multifaceted projects come with a high level of complexity is crucial for successful project management and execution;
- embracing the need for flexibility and adaptability is essential in today's fast-paced and ever-changing world;
- adopting a holistic perspective when developing strategies, projects, or solutions involves considering the interconnectedness of various components and stakeholders within the system;
- instead of trying to eliminate uncertainty entirely, they should focus on building resilience and agility to thrive in uncertain environments, turning challenges into opportunities for growth and innovation;

DOI: 10.1201/9781003688334-1

- prioritising open communication and collaboration among team members and stakeholders is essential for fostering a positive work environment and achieving project success;
- cultivating a mindset of continuous learning and improvement is essential for personal and professional growth; and
- taking a long-term view when managing these multifaceted projects to navigate the complexities of multifaceted projects while ensuring that they contribute positively to the long-term goals of the organisation.

By embracing these methodologies, project managers can more effectively navigate the complexities of multifaceted projects and enhance the likelihood of success. Additionally, balancing a focus on the big picture with attention to intricate details is crucial for achieving success. Westland (2007; p xv) states, "*Why such a large percentage of projects fail to deliver, typical causes include poor project sponsorship, undefined requirements and miscommunication.*" Effective leadership is crucial for setting a clear vision, aligning team goals, and providing the necessary support and resources to ensure successful project execution. This not only sets a clear path for success but also cultivates an environment where team members feel valued, supported, and motivated to perform at their best.

Mega LNG projects are ambitious undertakings marked by technical intricacies, large-scale operations, stringent regulatory requirements, and substantial financial investments. Inexperienced project managers face a variety of challenges that can be daunting when entering this arena that demand adaptive learning, proactive management approaches, and a deep understanding of industry-specific dynamics. These challenges in a project context can lead to various negative outcomes. Missteps can occur due to misunderstandings, lack of communication, or insufficient planning. Delays may arise from unforeseen obstacles, resource shortages, or dependencies on external factors. These issues not only disrupt timelines but can also lead to increased costs, as organisations may need to allocate additional resources to address problems or expedite processes.

Additionally, neglecting the human aspect of project management can have significant repercussions. This can lead to communication breakdowns, conflicts, and morale issues, all of which can negatively impact project performance. By prioritising the people involved in mega LNG projects and recognising their importance to overall project success, organisations can foster a collaborative and productive work environment that supports effective project delivery and mitigates potential challenges associated with communication breakdowns, conflicts, and morale issues.

By emphasising the importance of leadership, investing in project management expertise, and prioritising the well-being and collaboration of project teams, organisations can mitigate these challenges and enhance project outcomes. This holistic approach not only drives project success but also contributes to a more engaged and resilient workforce.

This book can play a vital role in highlighting these issues and offering guidance on how organisations can enhance their project management practices to address them effectively.

Addressing these challenges requires a complete methodology that emphasises leadership development, talent management, and fostering a people-centric project culture. By investing in capable leadership, providing adequate support and training for project managers, and fostering a culture of collaboration and empathy, organisations can mitigate risks and enhance the success of mega LNG projects. It is proposed that mega complex LNG projects cannot be managed using the iron triangle approach alone as these are often too complex and multifaceted in nature. These projects involve numerous intricate factors, including technological complexities, regulatory compliance, environmental considerations, stakeholder management, geopolitical factors, and supply chain intricacies. Managing such projects solely through the lens of the iron triangle can overlook these critical aspects. Westland (2007; p 4) states, ". . . organisations still don't really understand project management, despite having shelves full of textbooks. They think that project management is about the triple constraint's triangle."

Although these multidimensional projects are benchmarked throughout their project life-cycles, using skilled benchmarking houses, challenges in meeting budget, and schedule targets persist. This could be contributed to the lack of updated Scope of Work (SoW) information communicated to benchmarking houses. Such gaps can contribute to challenges in effectively managing mega LNG projects, as some organisations may believe there is no need to revisit the benchmarking data periodically. Communication gaps can lead to significant misunderstandings and discrepancies in benchmarks.

Addressing the challenges associated with outdated SoW information in mega LNG projects requires a comprehensive approach that goes beyond benchmarking alone. Ensuring that the SoW is regularly updated and shared with all relevant stakeholders, including benchmarking houses, is crucial for maintaining accuracy and alignment. This can help avoid any assumptions or outdated data influencing the benchmarking process.

By adopting a thorough methodology that incorporates various elements and perspectives into a solution process, mega LNG projects can effectively address the challenges associated with outdated SoW information. This approach, by leveraging streamlined communication channels and standardised processes, not only enhances project efficiency and cost-effectiveness but also reduces the risks associated with miscommunication and discrepancies in project execution. The successful execution of mega LNG projects necessitates a holistic approach that includes precise SoW documentation, proactive communication, robust risk management, and active stakeholder engagement. By integrating these factors, projects can achieve better efficiency, mitigate risks, and foster effective collaboration among all parties involved. In the context of mega LNG projects, relying solely on benchmarking data to manage the SoW effectively can be limiting. Benchmarking provides valuable reference points, but it may not fully address the unique complexities and specificities of mega LNG projects. A more comprehensive approach that integrates tailored SoW management practices alongside benchmarking data is essential for achieving optimal project outcomes. While benchmarking provides effective awareness into industry standards and best practices, it often falls short in addressing the dynamic and specific needs of individual projects. It involves effective risk management, proactive mitigation strategies, strong leadership, robust project governance (including corporate governance), and a focus on fostering collaboration and alignment among all project stakeholders. This holistic approach not only supports effective project delivery but also enhances resilience and adaptability in a dynamic and challenging industry environment. While benchmarking can offer valuable insights, it is merely one tool in the project management toolkit. It identifies that effectiveness is contingent upon how well it is integrated with broader project management practices. The evolving nature of massive LNG projects poses significant challenges in obtaining quantifiable benchmarking data for schedule activity durations and costs, especially during the Develop and Execute phases. This chapter explores the unique dynamics of benchmarking in LNG projects, examines current limitations, and proposes strategies to enhance benchmarking practices. The aim is to support informed decision-making and optimise project outcomes. By addressing the unique dynamics, overcoming current limitations, and implementing strategic enhancements, benchmarking practices in LNG projects can evolve to better support informed decision-making, optimise project outcomes, and drive industry-wide improvements in performance and efficiency.

1.1 Oil and Gas

When organisations use the term 'oil and gas', they typically refer to activities involved in the exploration, development, production, and distribution of petroleum products, encompassing both crude oil and natural gas. This whole process includes both crude oil and natural gas, which are critical resources for energy production and various industrial applications. This industry encompasses several stages, including prospecting for oil and gas reserves, drilling wells, extracting the resources, refining them into usable products such as gasoline or natural gas, and transporting these products to consumers or other industries. Accordingly, the term 'oil and gas' encompasses a broad array of activities within the energy sector. Abdel-Aal (2021; p 3) states, "The oil & gas industry is one of the largest sectors in the world in terms of dollar value, generating an estimated $3.3 trillion in revenue annually." Although natural gas is a vital component of the industry, it often receives less attention in public discourse compared to oil. This is despite its growing significance as a cleaner alternative to coal, it plays a crucial role in electricity generation, heating, and industrial applications. However, with the increasing emphasis on sustainability and the transition to cleaner energy sources, natural gas is gaining greater prominence in discussions about the future of the energy industry.

When a company aims to explore and potentially extract oil and gas resources in a specific area, it generally must obtain exploration permits or licenses from the government of the country where the resources are located. Governments play a crucial role in managing their country's natural resources, ensuring that exploration and production activities are conducted responsibly and in compliance with national laws and international best practices. This regulatory framework helps to balance the interests of energy development with environmental protection and economic benefits for the country. Exploration permits in the oil and gas industry are typically granted for fixed durations, which can vary depending on the country and its regulatory framework. These permits come with various obligations and commitments for the lessee (the company holding the permit). These obligations are intended to ensure that exploration activities are conducted effectively and responsibly within the designated area throughout the permit's validity period. Once an exploration permit is approved, the company holding the permit usually makes significant investments to conduct activities aimed at identifying potential oil and gas reservoirs. One of the primary methods used in this phase is seismic surveys. These surveys involve specialised equipment to generate and record sound waves that travel through the Earth's subsurface. These waves bounce off rock formations and other geological structures, providing data that can be analysed to create detailed images of underground layers.

Seismic surveys are a crucial part of the exploration process in the oil and gas industry, particularly in offshore areas where visual inspection is impossible due to the water depth. By analysing the data and images obtained from seismic surveys, geologists and geophysicists can identify potential hydrocarbon reservoirs and evaluate various characteristics such as size, shape, and geological properties. It provides a foundation of knowledge that guides decision-making on drilling locations, resource potential, and subsequent development strategies. They play a key role in reducing exploration risk and maximising the chances of successful resource discovery in the oil and gas industry.

If further investigations indicate that the reserves are economically feasible for development, the next step involves undertaking a drilling program to conduct exploratory wells. Seismic surveys are a crucial part of the exploration process in the oil and gas industry, particularly in offshore areas where visual inspection is impossible due to the water depth. By analysing the data and images obtained from seismic surveys, geologists and geophysicists can identify potential hydrocarbon reservoirs and evaluate various characteristics such as size, shape, and geological properties. It provides a foundation of knowledge that guides decision-making on drilling locations, resource potential, and subsequent development strategies. Herkenhoff (2018) points out that once the field is delineated, production wells are drilled – these are also referred to as development wells. Production wells serve as critical indicators for companies and joint venture partners (JVPs) to gauge the commercial viability of a field. By extracting hydrocarbons and measuring the flow rates, pressure, and quality of the produced fluids, they provide valuable data essential for assessing the field's potential for commercial investment. This data helps stakeholders make informed decisions about whether to proceed with development projects and allocate resources accordingly. Habibullah et al. (2009; p 10) state, "Reservoir size must support the plant capacity through the duration of the project life."

1.1.1 Natural Gas (NG)/Liquified Natural Gas (LNG)

Natural gas (NG) is not a new product, and it has been utilised as a commercial gas for centuries, particularly for lighting streets and homes before the advent of electricity. Natural gas has a rich history as a commercial fuel, evolving from the early use in ancient times to becoming a cornerstone of modern energy systems worldwide. The continued importance arises from the efficiency, lower carbon emissions compared to other fossil fuels, and the role as a transitional energy source in the global shift towards cleaner energy technologies. Meza et al. (2022; p 1) state "NG is reaching a global trade using liquefied natural gas (LNG), connecting importing, and exporting customers separated by oceans and continental distances." The first recorded commercial use of natural gas dates back approximately 2,400 years ago in ancient China. During that time, natural gas was extracted from shallow wells and transported through bamboo pipes. The gas was used primarily for heating in the production of salt from brine in gas-fired evaporators (Kidnay et al. 2011). This historical use of natural gas by ancient Chinese civilisations demonstrates early recognition of its utility as a fuel for industrial processes. The extraction and use of natural gas in such applications marked an important milestone in the commercialisation of this resource. Since then, natural gas has continued to evolve its usage and importance, becoming a key component of global energy systems due to its versatility, relatively cleaner burning properties, and widespread availability.

Historically, natural gas was often considered a byproduct of oil extraction and was sometimes seen as a nuisance rather than a valuable resource. In many oil fields, natural gas was flared off or vented because the infrastructure and technology needed to capture, process, and transport it were either lacking or deemed too costly. According to Tusiani and Shearer (2007), historically these gas reserves were considered as something that impeded the development of oil reserves and viewed as a nuisance. Today, natural gas plays a crucial role in global energy supply, providing a significant portion of the world's electricity and serving as a transitional fuel in the shift towards cleaner energy sources. The historical perspective highlights how perceptions and utilisation of natural gas have evolved from being a byproduct and nuisance to being recognised as an important component of the energy mix. Rahmat et al. (2022; p 1) state, "In the race towards greener energy, Liquified Natural Gas (LNG) stands out as an excellent alternative to reduce greenhouse gas emissions and help combat global warming."

1.1.2 LNG in Australia

The LNG industry in Australia was relatively minor compared to other major LNG producers like USA, Qatar, Russia, and Indonesia, until the discovery of large reserves of natural gas, condensate, and LNG in the northwest of the country, particularly in the Browse basin, Carnarvon basin, and the Bonaparte basin, that marked a turning point. The development of these reserves, along with substantial investments in liquefaction plants and export infrastructure, led to a surge in Australia's LNG industry and transformed Australia into a major player in the global

LNG industry. This growth underscores Australia's strategic position as a reliable supplier of LNG to energy markets worldwide. These fields were discovered in the northwest shelf in the 1970s but for many years lay dormant as they were too expensive to put into production.

Cassidy and Kosev (2015) suggest that LNG would become Australia's second-largest commodity export.

According to Twomey (2012), there were US $185 billion worth of oil and gas projects under construction or at the advanced stages of planning and approval, and they represent over a third of all business investment. Awwad et al. (2021; p 1) state, "Since the turn of the century CCS has been held up as a way in which coal and gas could go on being used to generate electricity without wrecking the climate, both by being retrofitted to existing plants and designed into new ones."

Australia has made remarkable strides in the liquefied natural gas (LNG) industry over the years. With the abundant natural gas resources and strategic investments in LNG infrastructure and technology, Australia has positioned itself as a major player in the global LNG market.

Australia's LNG exports were forecast to ease to 81.2 Mt in 2022–23 from 83.2 Mt in 2021–22, as field depletion issues at Darwin LNG offset higher utilisation rates at other Australian LNG facilities (Figure 1.2). Volumes were forecast to stabilise at 80.9 Mt in 2023–24, with no significant changes expected until the Northwest Shelf depletes (and a second train at Pluto LNG comes online) beyond the outlook period.

Qatar has long been the leading exporter, but Australia's rapid growth in LNG production has enabled it to secure a strong foothold in the industry, making it the

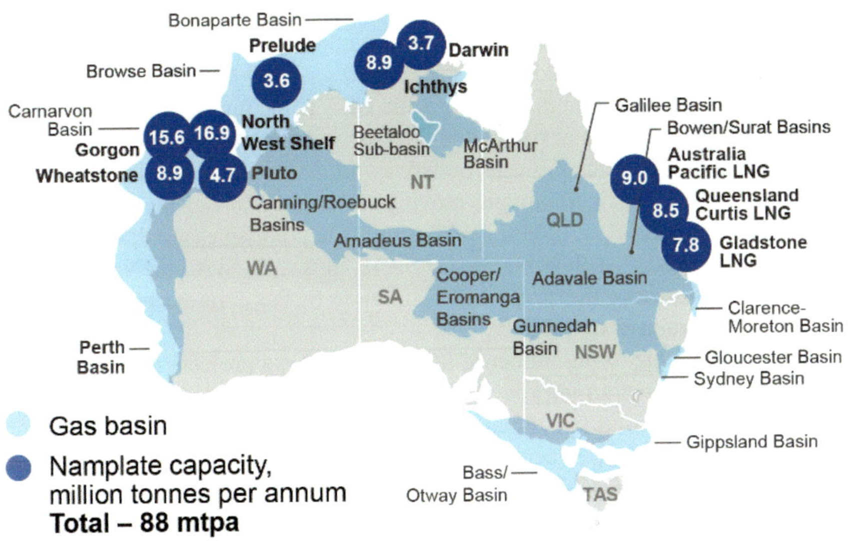

Figure 1.1 Number of LNG Projects and Gas Basins in 2022

Source: Australian Government Department of Industry, Science and Resources 2023.

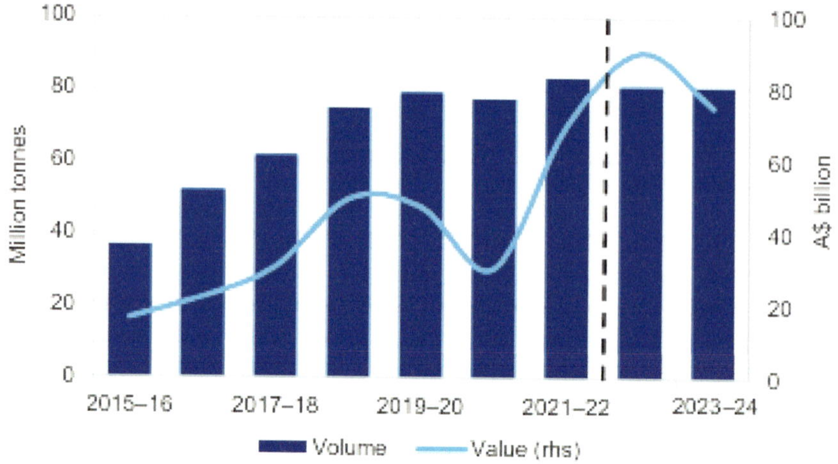

Figure 1.2 Australia's LNG Exports by Value and Volume (2022)

Source: ABS (2022) International Trade in Goods and Services. 5368; Department of Industry, Science and Resources (2023).

second-largest exporter globally after the USA. This development reflects Australia's growing importance in the energy sector and the ability to capitalise on natural resources to meet global energy demands.

LNG is becoming a crucial feature of the northwest of Australia with ConocoPhillips' Bayu Undan on-stream, Woodside's NWS on-stream, Woodside's Pluto on-stream, and Chevron's Gorgon and Wheatstone on-stream. In 2013, Sakmar suggested that Australia would become the world's second-largest exporter of LNG behind Qatar. As illustrated in Figure 1.3 the USA is the highest exporter of LNG followed by Australia. Qatar is now the third-largest exporter of LNG.

There are more LNG projects underway in Australia, such as Woodside's Browse, Sunrise, and Scarborough, moving towards Financial Investment Decision (FID). Cameron et al. (2005; p 686) state, "Oil companies are developing gas fields in remote areas, such as Northwest Shelf Australia and Arctic Norway."

Table 1.1 illustrates the status and magnitude of the current and future LNG projects in Australia as of 2022.

Accordingly, due to the immense investments these projects require, organisations need to adopt an improved form of the Performance Management Systems (PMS), including benchmarking, to ensure these projects are delivered on time and on budget. Weems and Hwang (2013; p 270) state, "Projects in Australia, Papua New Guinea, Norway, Indonesia (Tangguh) and Alaska followed the 'Non-Incorporated Joint Venture' structure approach. For example, each participant in the Northwest Shelf Project in Australia owns one-sixth of the LNG plant, supplies one-sixth of the feed gas and is entitled to one-sixth of the revenues."

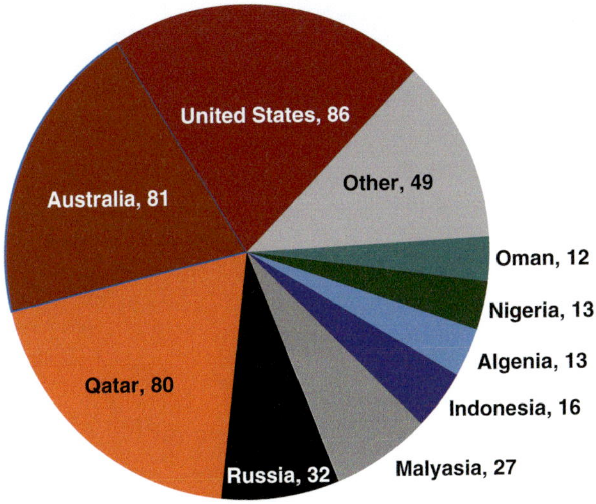

Figure 1.3 Global LNG Exports, 2023

Source: Institute for Energy Economics and Financial Analysis (2023; p 11)

Table 1.1 Australia LNG Projects 2022 Status

Australian LNG Projects as at 2022 Status			
Company	*Develop Phase*	*Execute Phase*	*Operational*
ConocoPhillips			Darwin LNG
Woodside Energy			NWS LNG
Woodside Energy			Pluto LNG
Chevron			Gorgon LNG (BWI)
Chevron			Wheatstone LNG
INPEX			Ichthys LNG
Shell FLNG (Prelude)			FLNG (Prelude)
Santos	Bonaparte LNG		
Woodside Energy	Browse LNG		
Woodside Energy	Sunrise LNG		
LNG Limited	Fisherman Landing		
Woodside Energy		Scarborough LNG	

Source: Author's Research

Notwithstanding the north-western Australian LNG projects, there are other complex oil and gas projects undertaken around Australia, such as the Coal Seam Gas (CSG) projects in Eastern Australia, as shown in Figures 1.4 and 1.5.

Coal Seam Gas (CSG), also known as coal bed methane (CBM), is a natural gas extracted from coal seams. It primarily consists of methane and is found in the pores of coal deposits.

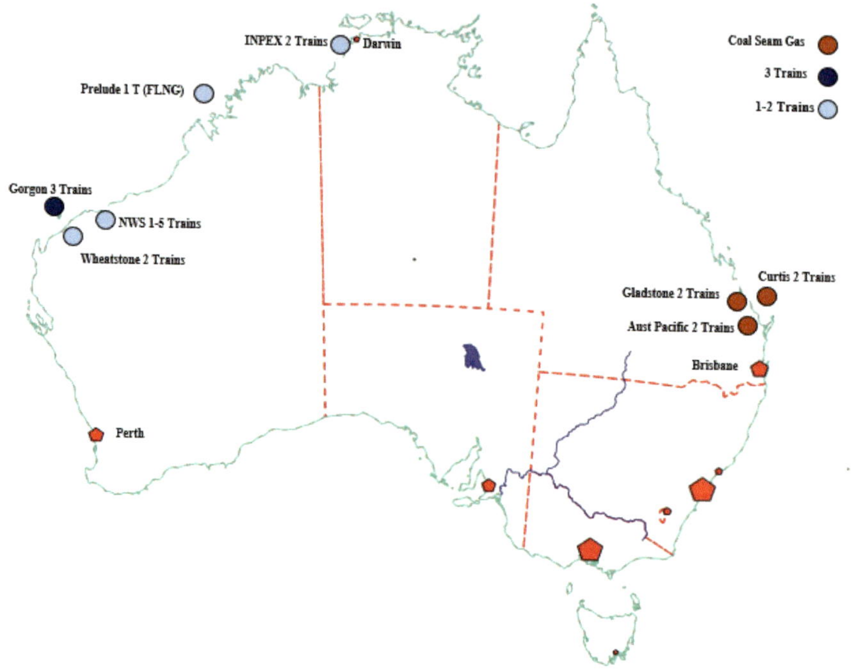

Figure 1.4 Gas Assets (Trains) in Australia
Source: Author's Research

1.1.3 LNG Value Chain

Shingan et al. (2024; p 430) state, "The entire process from natural gas extraction to consumption is covered in the presented analysis of the LNG value chain, starting out by looking at the discovery and production of natural gas reserves, highlighting important innovations and methods used during the extraction stage." According to Cameron et al. (2005), LNG is an important commodity with increased significance; however, it is complicated to produce, process, ship, and distribute. Therefore, the value chain is an important process within an organisation, and it needs to be scrutinised carefully during each sector to ensure there are no broken links that could disrupt the process. Shingan et al. (2024; p 441) state, "Operations that are efficient and economical are essential to the overall profit ability of the LNG value chain." The LNG value chain consists of onshore and offshore components which commence with the exploration, drilling/production (offshore) and gas processing/liquefaction (onshore) then transport to customers for degasification and sales. Smith and Herscowitz (2016; p 22) state, "The LNG value chain begins with the drilling and production of natural gas from subsurface gas reservoirs, usually offshore."

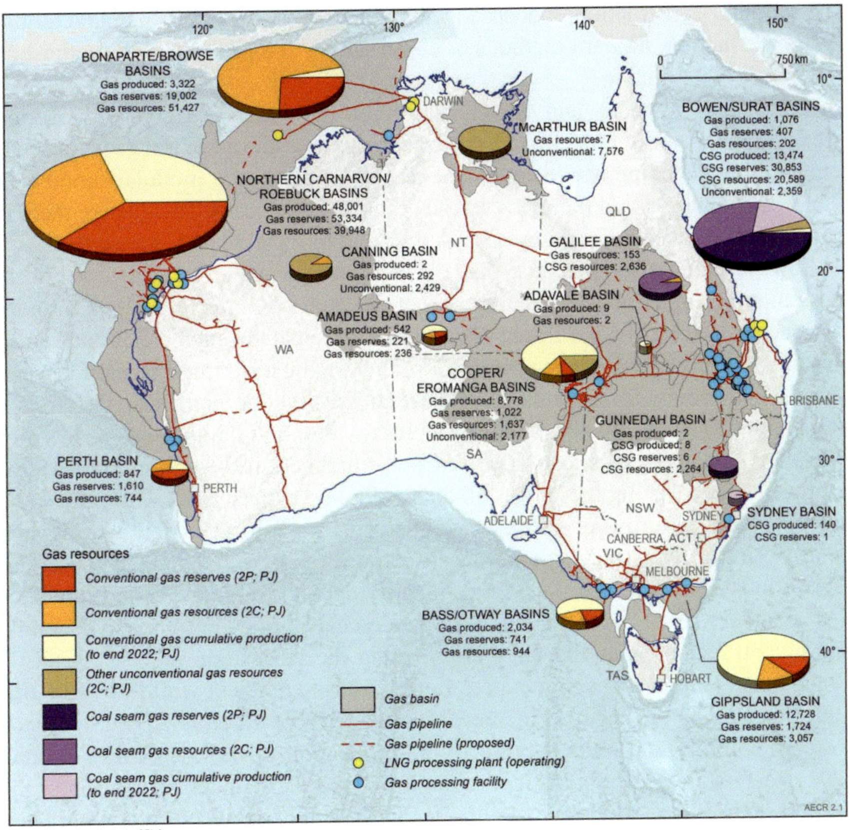

Sources: Geoscience Australia; GPInfo.
Pipeline routes from the GPInfo petroleum database.
Note: LNG = liquefied natural gas, PJ = Petajoules.

Figure 1.5 Location of Australia's Gas Resources and Infrastructure

Source: Geoscience Australia 2012

1.1.4 Downstream

Due to the size of the downstream portions of mega LNG projects, more organisations are using a modular approach to the fabrication of the plant, compared to stick-built. Phalen and Scotti (2009; p 41) state, "Modularization use in non-traditional areas such as LNG and mining construction projects, as well as the increasing demand from traditional projects, is placing a strain on fabrication yard capacity."

The process plants for downstream are also becoming cheaper to fabricate as modularised units overseas (rather than stick-built on site) and shipped from major fabrication yards to the plant sites for installation. Phalen and Scotti (2009; p 41) state, "Modularisation is an acknowledged strategy to combat the impact of construction labour shortages." Subsequently, Wuni and Shen (2020; p 3) state

". . . modular construction, industrialized building systems, offsite manufacturing, offsite production, modern methods of construction, and prefabricated prefinished volumetric construction."

Accordingly, Phalen and Scotti (2009) propose that using modularisation streamlines the work process and moves labour to a more controlled yard environment thus reducing onsite costs. The reason for opting for modularisation is to reduce the manning and the many interfaces on these massive sites where there is usually limited accommodation, access, transportation, and lay-down areas.

However, this also demands that the benchmarking timeframes to fabricate these modules must be realistic. At present, these are optimistic, leading to more work being pushed into the future periods than anticipated which creates an excessive amount of Carry-Over-Work (COW). This can be a common challenge when plans or projections don't match up with actual processes or realities on the ground. Managing COW effectively requires regular reviews and adjustments to the project plans, clear communication of realistic timelines, and a more conservative approach to forecasting these. It is all about aligning expectations with achievable outcomes and ensuring everyone is on the same page to avoid bottlenecks and delays. Setting realistic benchmarking timeframes is essential for the successful fabrication of modules. This involves identifying achievable goals and timelines that consider various factors such as production capabilities, resource availability, and potential challenges that may arise during the fabrication process. Realistic benchmarks help in managing expectations, ensuring efficient workflow, and ultimately achieving the desired outcomes within a feasible timeframe. This approach fosters reliability and sustainability in the production of these modules. Eftimie (2021; p 1) states "Modularization aims to reduce the number of interfaces, the Total Installed Cost (TIC) and overall schedule length of a project while optimizing the return on investment and allowing standardization of future similar projects." Ensuring that module drawings are free from 'HOLDS' during the bid stage or fabrication stage is crucial for cost-effectiveness and efficiency. When drawings have 'HOLDS', it typically means there are issues or uncertainties that need resolution, potentially leading to delays, rework, and increased costs during fabrication. By providing clear, accurate, and finalised drawings early on, this minimises the chances of costly changes or disruptions later in the fabrication process. This approach helps fabrication yards proceed smoothly with their work, adhering to the agreed timelines and budgets without unexpected expenses due to revisions or corrections. It is a proactive step that supports cost control and timely project delivery. Eftimie (2021; p 1) states, "Modularization offers many significant benefits, including cost savings by reducing field erection, higher quality and safety from having fabrication done in an offsite shop, decreased schedule, increased efficiency, module mobility and reusability, and less site construction complexity due to fewer interface points for modules, reduced onsite logistics, etc."

Projects such as Woodside's Pluto and Chevron's Gorgon can provide real tangible benchmarking data on downstream module fabrication durations, if shared without restraint amongst the industry. To address these challenges, it is essential for organisations undertaking mega LNG projects to prioritise comprehensive engineering design during the Develop phase. This includes engaging experienced engineering firms, conducting thorough feasibility studies, and involving key stakeholders early in the design process to ensure alignment with project objectives and requirements. Olugbenga et al. (2015; p 1) state, "There has been limited published research that has specifically examined the nature of cost overruns in mega-hydrocarbon projects."

These downstream mechanisms of the projects are becoming so large and complex, as train sizes increase due to the economy of scale, it becomes increasingly challenging to manage and coordinate all of the downstream components. This poses challenges for conventional benchmarking processes, which can struggle to accurately assess large-scale components due to their complexity and unique issues and may not be well-suited to accurately assess such large-scale projects. To simplify the way the downstream, execute phase is benchmarked, breaking down this phase into three specific areas – Infrastructure, Causeway/Jetty, and LNG plant – can simplify the process and provide more focused insights. It is believed that mega complex LNG projects characterised by their scale, complexity, and technological demands, often exceed the capabilities of project management techniques. Ma and Fu (2020; p 2430) state, "The starting point for the complexity management of mega construction projects involves understanding the project complexity and the corresponding effects on the implementation and eventual deliverables of the projects." Whereas Priemus et al. (2008) suggested that mega projects with their ever-growing complexity may not be manageable using traditional project management processes/techniques. The complexity involved in developing LNG projects necessitates a more sophisticated and integrated approach to project management. Identifying and addressing the potential lack of local qualified skill sets and the reluctance to pre-invest in infrastructure are crucial aspects that should be managed early in the project life-cycle for LNG projects. By proactively addressing these challenges early in the project life-cycle, LNG project managers can better manage risks, optimise resource allocation, and enhance project feasibility and success. This holistic approach not only supports project execution but also contributes to sustainable development and positive outcomes for all stakeholders involved. According to Kerzner (2006), existing project management methodologies were developed to manage small projects and, as such, cannot effectively support mega projects. The development of massive projects like LNG facilities demands a highly skilled and diverse team to effectively address the myriad challenges associated with scale, complexity, and specialisation. This is also because the development of massive projects such as LNG facilities necessitates a much larger and more diverse array of skilled personnel compared to smaller-scale projects. This can make the use of current benchmarking techniques obsolete when undertaking mega complex LNG projects, which can appear to be an antiquated tool used to tick a box in the life-cycle process of project approvals.

With the more common smaller-type projects, the current project management techniques used are adequate to develop these projects. However, when dealing with mega complex LNG projects, it is necessary to work across or interface between many diverse levels and locations (local and international) of a project throughout the project life-cycle as they are extraordinarily complex and multifaceted developments. Love et al. (2015) suggest that project complexity is the interdependence which can exist between numerous interrelated elements within a project. Furthermore, being so large and multifaceted, they are separated into modules and, as such, are fabricated in numerous locations throughout the world. This is when it is imperative that a compelling case should be made to have Interface Management (IM) at the highest levels within the project management portfolio. IM plays a critical role in the success of large-scale projects, especially in complex endeavours like LNG facilities. Nooteboom (2004) points out that although IM is not new to project management, it should become a critical project component, although this has not been fully appreciated or appropriately addressed for mega projects.

According to Nooteboom (2004), it appears that the management of interfaces on complex projects, referring to common boundaries such as people, systems, equipment, concepts, and locations, has been a hidden aspect of project management within mega projects. Therefore, it could be argued that, if the current project management processes were to increase the focus on developing IM interaction with these multifaceted LNG projects, this would enhance transparency across the life-cycle of these difficult projects. Accordingly, as the scale and scope of complex LNG projects become more visible during each phase of development, phase-based benchmarking can significantly enhance control and measurement. For that reason, phase-based benchmarking enhances control over complex LNG projects by providing a structured framework to monitor progress, manage risks, optimise resources, and ensure quality throughout the project life-cycle. It enables project teams to navigate the inherent challenges of scale and scope more effectively, leading to more achievable and realistic project outcomes. Performance metrics are an essential part of benchmarking and are the key to effective measurements to evaluate various performance outcomes, Yun et al. (2016).

Oakland (2007) discussed how benchmarking is part of the TQM process and, therefore, is part of the TQM tool kit. It could be debated that it is time to revisit whether the benchmarking for complex projects like LNG developments should be treated as a distinct entity rather than just a component of Total Quality Management (TQM) toolkits. However, while there are merits to considering benchmarking as a distinct entity for managing complex projects, careful consideration should be given to integration with overall quality management frameworks, alignment with project management processes, and strategic objectives. The decision would depend on the specific needs, complexities, and goals of the organisations/projects.

It is, therefore, recommended that benchmarking should be a distinctive process tailored to the unique needs and complexities of each project for providing development metrics when undertaking these multifaceted LNG projects. A development metric can be an early warning indicator allowing managers to efficiently measure project performance at the phase level, which would then provide meaningful comparisons and effective strategies for improving performance in subsequent phases (Yun et al. 2016). Therefore, one could assume that, by doing so, more emphasis is focused on making the benchmarking process more relevant to the complexity of these large LNG projects, rather than being an item in a suite of other tools.

Benchmarking is far more than just a performance measurement tool; it is a critical process that provides the structure for effective governance throughout the entire life-cycle of an LNG project. Benchmarking helps to set clear performance expectations, providing a concrete reference point that helps guide the project team through every phase of the project life-cycle. It stands out as a prevalent process that would require a robust governance system when developing each of the phases within the project life-cycles of these complex LNG projects. Priemus et al. (2008) also discussed the governance regimes of mega projects and how they should comprise processes and systems needed to ensure successful investments and correct decision-making processes. Formal decision-making procedures do not reflect the reality of mega projects' sharp fluctuations, complexity, and unanticipated changes because of factors such as the political environment and technological novelty they process (Arena and Molloy 2010).

Arena and Molloy (2010; p 5) state, "In contrast with other governance forms, megaprojects are characterised by temporary cooperation between multiparty contractual actors." The process they have described deals mostly with the front-end governance of public mega projects. Drawing on the outcomes from surveys, interviews, and informal discussions from those who have worked on several mega LNG projects, this governance process is far from adequate in allowing stakeholders of these complex LNG projects to be comfortable in investment within the projects, especially as most of these mega projects have overrun on budget and schedule. Flyvbjerg et al. (2003) discuss the problem of the lack of accountability and risk negligence as part of a governance process that does not really exist within current mega projects. Therefore, it could be assumed that without accountability there is no reason to benchmark these multifaceted projects with quantifiable data due to lack of ownership to the success of the project. However, accountability is crucial for effective benchmarking of multifaceted projects like mega LNG developments. Therefore, fostering a culture of accountability within organisations undertaking multifaceted projects is essential for ensuring that benchmarking efforts are effective and impactful. It encourages proactive management, fostering collaboration, improved communication and efficiency among team members, and enhances overall project performance from inception to completion.

According to Andersen and Pettersen (1996), benchmarking measures is one of the many tools for improvement of a company's critical processes. They reflect on the use of models such as the benchmarking wheel where the process is set as Plan, Search, Observe, Analyse, and Adapt. This process allows companies to recognise and establish the critical success factors to identify what and how they are going to benchmark, or if they take the benchmarking processes as a means of creating a successful project. Alternatively, whether they merely make a tokenistic effort to the process. Other process models are those as described by Oakland (2007), which outlines the new framework for Total Quality Management (TQM). This is a four-step process shown as Planning, Performance, Process, and People as illustrated in Figure 1.6. This provides a structured framework for managing and improving complex projects like LNG developments.

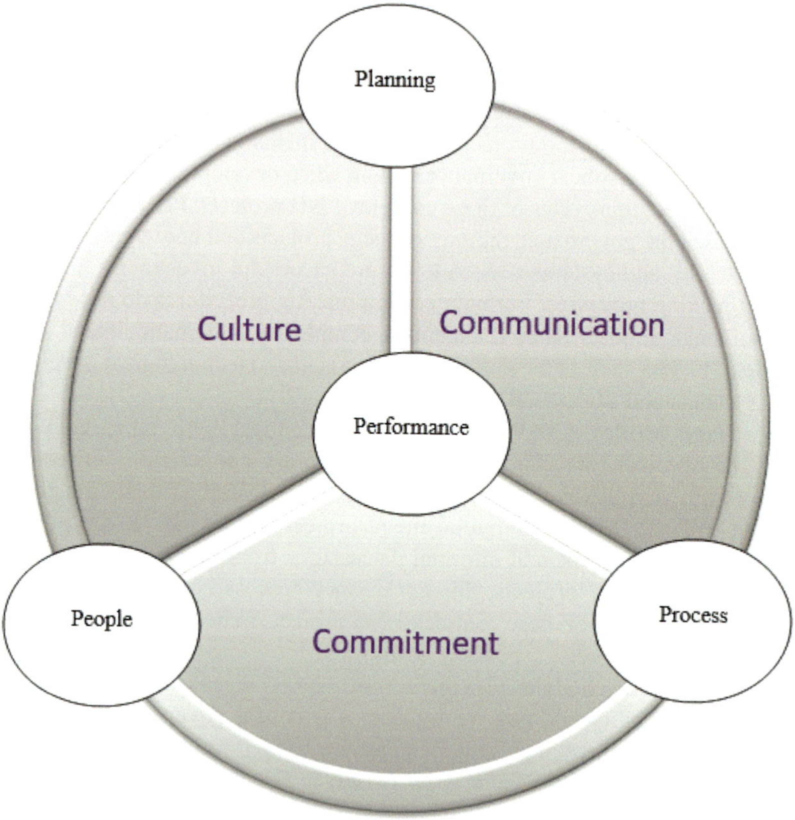

Figure 1.6 The New Framework for Total Quality Management
Reformatted from Oakland (2007; p 27)

1.2 What Is a Project Life-Cycle?

All organisations, books, journals, etc., use different descriptions for the project life-cycle; however, for the purposes of this book, the five phases are described below for developing mega complex LNG projects. A project life-cycle is a structured sequence of phases or stages a project goes through from its initiation to its completion and close-out. It provides a constructed approach to manage a project and ensures that all necessary steps are taken to deliver a successful outcome.

The project life-cycle model consists of a series of distinct, sequential phases that a project goes through throughout its life-cycle. For instance, Assess (initiate), Select (pick), Develop (mature), Execute (activate), and Operate/Evaluate (appraise/project-close-out). By dividing the project into these distinct phases, this provides several advantages for project managers in effectively managing resources, mitigating risks, and ensuring smooth progression from initiation to closure. Subsequently, this facilitates effective management of resources, risk mitigation, and smooth project progression. It supports project managers in achieving project objectives while ensuring stakeholders are engaged and quality standards are met throughout the project life-cycle. It supports project managers in achieving project objectives while ensuring stakeholders are engaged and quality standards are met throughout the project life-cycle.

Each phase has its own separate set of deliverables, activities, and milestones, contributing to the overall success of the project. Each of these phases represents a different stage in the project's development and execution. In brief each phase is described below:

- Assess (initiate):
  - This is the first phase where the project's feasibility, scope, and objectives are evaluated. It involves identifying the need for the project, understanding the goals, and determining whether it aligns with the organisation's strategic objectives. This phase lays the foundation for the entire project.
- Select (pick):
  - In this phase, the project's specific approach and resources are determined. Project managers and stakeholders decide on the project's scope, timeline, and budget. Additionally, a project team is assembled, and responsibilities are assigned to team members.
- Develop (mature):
  - FID is an important step in the project management and investment process, to move the project into the execute phase;
  - This is a crucial phase where detailed planning, design, and development occur. The project's feasibility is thoroughly assessed, considering technical, financial, and operational aspects. If the project passes feasibility tests, it moves forward with detailed design and implementation planning.
- Execute (activate):
  - In the execution phase, the project plan is put into action. This involves coordinating the various tasks, activities, and resources required for project implementation. Continuous communication, regular monitoring and control are vital to ensure that the project stays on track and any deviations are addressed promptly.
- Operate/Evaluate (appraise/close-out):
  - Once the project is complete and operational, this phase focuses on evaluating the performance and outcomes. This assessment helps identify whether the project's objectives were met, and it provides insights for continuous improvement. Lessons learnt from the project are captured and applied to future projects;
  - This is where the Capital Expenditure (CAPEX) is closed, and the transition from CAPEX to operations is critical to ensuring that the project achieves it intended objectives and provides a return on investment. The project is

handed over to the operations team, it generally marks the transition from the project's active development and implementation phase to its ongoing maintenance and support phase. This is a critical step that involves the coordination between various teams, including project management, engineering, finance, and operations, to ensure a smooth handover and the seamless operation of the asset or facility. This phase is often referred to as the "commissioning" or "project handover" phase.

The aforementioned structure is a typical representation of a project life-cycle model. Each phase is designed to address specific challenges and tasks related to different stages of the project's evolution. It is important to note that while this structure is common, the specifics can vary based on the nature of the project, the industry, and the organisation's practices. The project life-cycle model serves as a roadmap to guide project managers and stakeholders through the entire project journey. These will be discussed in-depth in Chapter 2 – Project Life-cycle.

The project commences with the Assess phase, and it is also known as the initiation phase. This is when the project begins developing. During the Assess (Initiate) phase, the project begins with the development of the main strategic documents:

- Project Charter;
- Project Management Plan (PMP);
- Project Execution Plan (PEP);
- Project Management Systems (PMS);
- Key Performance Indicators (KPIs);
- Balanced Scorecards (BSCs); and
- Level 1 budgets and schedules.

These documents are the basis of the overall project life-cycle and are live documents that ought to be revised at the end of each phase and before the risk gate phase approval to continue into the following phase.

At the Assess phase, the 'Phase-Based Benchmarking' (PBB) process is introduced. The PBB looks at the higher level of benchmarking the project life-cycle, which evaluates project performance by each phase and establishes initiative-taking strategies for improving performance in each phase for measuring the subsequent phases (Yun et al. 2016). This approach involves evaluating how well a project performs in each individual phase and using the insights gained to develop strategies for enhancing performance in subsequent phases. The PBB is when the organisation/project management team develop a higher level of measurement which is the performance measurement or benchmarking of each phase of the project life-cycle. This process is discussed in-depth in other chapters in this book.

Each phase has separate deliverables, and in some cases a different project manager. However, during the course of each chapter of this book, it will establish if there needs to be a mindset change to discuss alternative options for phase durations, especially the Develop and Execute phases.

1.3 What Is a Mega/Giga Complex LNG Project

There are many types of projects from small, medium, large, and very large. In this book the analysis will examine mega complex LNG projects and a need for some of these multifaceted projects change to a more apt descriptive of giga, as they become more complex.

All projects are unique in nature, and all have defined timescales with start and finish dates with specified deliverables, both client and project specific requirements. For clarity, it is proposed to divide these mega complex LNG projects into four distinct categories, i.e., small conventional LNG projects, large conventional LNG projects, large complex LNG projects, and exceptionally large/extraordinarily complex LNG projects. By doing so, this provides specific benchmark constructs for organisations for each phase and overall life-cycles. The following classifications, due to the economy of scale, and complexity, are recommended:

- **Small Conventional LNG Project**
  Not complex + small + under US $ 1 billion. These projects typically involve lower capacities and less intricate configurations. They are well-suited for serving local markets or niche applications.
- **Large Conventional LNG projects**
  Not complex + large + over US $1 billion under US $10 billion. Moving up the complexity ladder, large conventional LNG projects involve greater capacities and extended supply chains. The distinction between large and small projects enables organisations to allocate resources effectively, account for increased regulatory considerations, and implement supply chain optimisations to meet broader market demands.
- **Large Complex LNG Projects**
  Complex + very large + over US $10 billion and under US $20 billion = mega project.
  (Example: Pluto LNG Project (CAPEX) ~US $13 billion (large = single train). This category recognises projects that deviate from the standard large-scale projects due to their intricacy. Organisations can take this distinction as an indication to incorporate advanced risk management strategies, enhanced project coordination, and specialised engineering
- **Exceptionally Large/Extraordinarily Complex LNG Projects** (Very Large/Very Complex LNG Projects).
  Extraordinarily complex + very large + over US $20 billion = giga project.
  (Example: Gorgon LNG Project (CAPEX) ~US $54 billion (economy of scale = three trains).

At the pinnacle of complexity, this category recognises LNG projects that are unparalleled in their scope, technological demands, and challenges. These projects typically involve intricate engineering, large-scale infrastructure development, and cutting-edge technologies. They may also be set in remote or challenging locations, requiring innovative solutions to overcome environmental, logistical, and regulatory hurdles.

Organisations can prepare for these ventures by fostering collaboration among multidisciplinary teams, involving top-tier engineering partners, and devising innovative solutions to address intricate engineering, regulatory, and logistical hurdles.

Mega complex LNG projects are multifaceted in their nature as their size increases as their complexity grows. The objective is to understand why mega/giga complex LNG projects have unpredictable consequences as they enter the Execute phase.

Mega LNG projects are some of the most complex and capital-intensive undertakings in the energy sector. Given their massive scale, technical complexity, and the diverse range of engineering disciplines required, it is virtually impossible for a single design house to manage the entire project. Therefore, these are divided into sections, and the design is farmed out to different overseas design houses adding an even further level of complexity to manage these multidimensional LNG projects. Careful planning, clear communication, and tight integration are crucial to managing this complexity.

As the mega complex LNG projects increase, it should be considered that a better descriptor for these is giga complex LNG projects. To identify whether change from mega to giga LNG project descriptor is required, a giga project would have a CAPEX of USD $20 billion or greater. Giga projects, by their very nature, are massive undertakings that typically involve multiple interconnected mega projects within them and have many sections that in themselves are actually mega projects. Each section of a giga project could be a complex, large-scale endeavour on its own, requiring significant resources, time, and expertise.

1.4 Chapter Summary

By integrating academic knowledge and industry perspectives, it has been possible to draw on established theories, methodologies, and research that provide a deeper understanding of fundamental principles, models, and the broader context. Incorporating industry knowledge brings in practical insights and lessons learnt from real-world case studies, operational challenges, market dynamics, and evolving technologies. This blend helps bridge the gap between theoretical knowledge and the practical challenges that organisations face when executing large-scale LNG projects. The result is a more holistic and actionable insight into how such projects should be planned, executed, and managed. However, due to the complexity of these mega complex LNG projects, including the LNG value chain, this has revealed that there is a dearth of relevant literature available pertaining to developing these massive LNG projects throughout their life-cycle, to enable a hypothetical finding to support the literature analysis. Instead, it has been established from those with experience and knowledge in developing the massive LNG complex projects. However, organisations developing these multifaceted LNG projects do not necessarily have or use such information when developing their LNG projects for success.

Organisations need to clearly understand what the definition or size/complexity of the project they are undertaking, as illustrated in Chapter 4 – Figure 4.1. Project Complexity Framework. It is a very complicated process to determine where their projects actually fit within the complexity framework.

Furthermore, it is crucial that they establish appropriate Key Performance Indicators (KPIs) and Balanced Scorecards (BSCs) that are tailored to the project's size, scope, and complexity. This will help track progress, measure success, and align stakeholders with the project's objectives. Consequently, this highlights the importance of aligning performance metrics with the specific characteristics of the project to ensure that the right factors are being measured and managed effectively.

To do so will necessitate a change in mindset on how they manage these multifaceted projects which will allow organisations and project teams to understand the project complexities and the framework required for success of their projects.

It should be noted that there is a substantial amount of benchmarking data available relating to the offshore (upstream) portions of these LNG projects. Such data is obtainable through benchmarking forums such as the Performance Forum (PF) and the Independent Project Analysis (IPA) organisations. Both benchmarking organisations have developed robust databases from worldwide mega LNG projects. However, the onshore (downstream) data for these projects is still insufficient due to the reluctance of organisations to share this data and the small number of these complex LNG projects developed.

As benchmarking techniques have been improved and refined over time, most of these complex LNG projects are typically benchmarked (measured) against five main phases, which, in turn, form the project life-cycle. These phases of the life-cycle are the Assess, Select, Develop, Execute, and Operate/Evaluate. However, different organisations use different terminology, but the result is the same; the project life-cycle is broken down into phases and check gates or assurance risk gates which are applied at the end of each phase.

The definition of the term giga complex LNG project is defined in section 3.1.2, which proposes that these complex giga projects would have a CAPEX of ~US $20 billion or greater. Complexity in mega/giga LNG projects continues to be an area that needs to be understood at each phase of the project to ascertain what the main contributors are that determines when to rename an LNG project from one size to another. Such an undertaking in the early phase of the project requires a new type of strategy, which then enables the project managers and their teams to openly formulate how they need to measure and manage these complexity issues in the project they are undertaking. This could also include a level of commitment from the organisation and JVPs to review and agree on how they need to deal with these complexities that they may be faced with during the project.

This assessment, although comprehensive, suggests that organisations ought to understand how to benchmark these multifaceted LNG projects. It is suggested that the project team should take the time to do a very comprehensive review of the Develop phase to ascertain the full SoW and formulate if the benchmarking data on schedule and cost is robust to proceed into FID, not just an opinion based on tacit information.

The emphasis on a rigorous, data-driven approach in the Develop phase is crucial for ensuring that the project is on the right track before moving to the Final Investment Decision (FID). The aim is to avoid relying on incomplete or subjective information and instead use robust, validated data to ensure the project is on track and feasible before moving forward to the FID stage. This will help mitigate risks and improve the likelihood of project success by addressing potential issues early in the process.

2 Project Life-Cycle: Overview

2.0 Overview

The project life-cycle for mega/giga LNG projects involves a more complex process compared to smaller LNG projects, primarily due to their economy of scale and the intricacy of integrating advanced technologies for managing extensive logistical and regulatory requirements. The scale and complexity of mega/giga LNG projects bring both significant challenges and substantial opportunities. These challenges require strategic planning, innovative approaches, robust risk management, and effective stakeholder engagement. As the global energy landscape evolves, LNG projects play a crucial role in meeting energy demands sustainably while driving economic growth and technological advancement.

Effective project management, stakeholder engagement, and risk mitigation are essential throughout each phase of the project life-cycle to ensure the successful delivery of these high-value projects. This chapter will focus on the activities, deliverables, and tasks that are essential to complete each phase of the project's life-cycle. In Chapter 6, a thorough examination of the life-cycle model is conducted, which identifies the deliverables for each phase that must be completed for the project life-cycle. This in-depth review of the life-cycle model demonstrates how each phase builds on the previous one and contributes to the overall success of the project, which, in turn, can highlight the interconnected nature of the project life-cycle. Furthermore, it identifies the importance of effective communication and collaboration between various stakeholders throughout the project's development.

2.1 Chapter Introduction

Examining the project life-cycle of mega/giga LNG projects, the author's revealed that extending the duration of certain phases is often necessary to address the complexity and the economy of scale of these multifaceted LNG projects. The current life-cycle phase durations are acceptable for smaller projects but not for mega/giga complex LNG projects.

DOI: 10.1201/9781003688334-2

2.2 Background

Similar to other projects, the mega/giga complex LNG projects have distinct phases that build up the project life-cycle. As identified in Chapter 1 and discussed in other chapters in this book, the five phases of the project life-cycle are a framework used to structure and manage the various phases of a project. Each phase builds upon the previous one, and although the phases are sequential, there can be overlap or iteration depending on the nature of the project and the project management methodology being used. Breaking down the project life-cycle into specific phases is crucial for effective project management, especially for complex and large-scale projects such as mega/giga LNG developments. This phased approach provides several key advantages that enhance both project understanding and management. While different organisations may use varied terminology for the phases of a project, the fundamental concept remains the same; a project typically comprises multiple phases, each with its own set of objectives, activities, and deliverables. Consequently, the concept of dividing the project life-cycle into distinct phases remains a fundamental principle in project management. This phased approach helps managing complex projects more effectively, ensuring that each aspect of the project is thoroughly planned, executed, and controlled. The phased approach to project management is instrumental in identifying and addressing issues early, which can significantly reduce the risk of major problems emerging later in the project life-cycle. Additionally, the review or approval process at the end of each phase is a critical component of project management. It serves several important functions in ensuring that the project remains on track and meets its objectives before proceeding to the next stage. Phase-end reviews and approval processes are integral for effective project management, providing a structured approach to verify that each phase meets its objectives before advancing. These processes ensure quality assurance, risk management, scope and budget control, stakeholder engagement, and informed decision-making. By incorporating formal reviews and approvals, projects can maintain control, address issues early, and increase the likelihood of successful project outcomes.

A structured approach to project management, with distinct phases and formal review and approval processes, significantly enhances clarity, control, and understanding throughout the project's progression. It provides clear goals and organised workflows, facilitates better monitoring and risk management, and ensures incremental learning and comprehensive oversight. This structured methodology not only improves project execution but also supports effective decision-making and stakeholder communication, ultimately contributing to the successful delivery of complex projects. While different organisations may use various descriptors for these phases, the core concept remains consistent, and the underlying principle of structuring the project life-cycle remains crucial for successful project management.

2.3 Oil and Gas Projects

The term 'oil and gas' often encompasses both the exploration and development of oil and gas resources. However, in practical usage within organisations and industries, the term can sometimes focus more heavily on specific aspects depending on

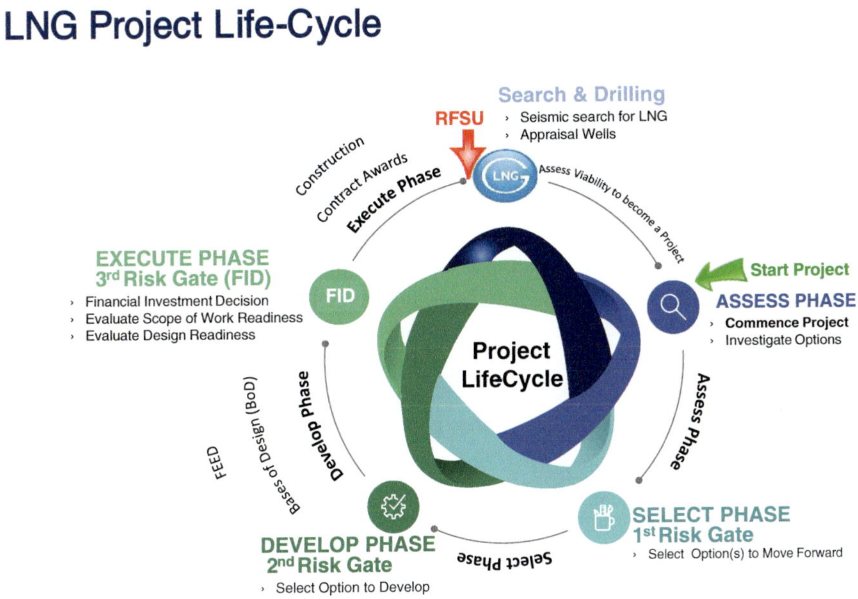

Figure 2.1 LNG Project Life-Cycle (From Search and Drilling to RFSU)

Source: Author's Research

the context. Figure 2.1 – LNG Project Life-Cycle defines the process and the three main risk gates that need to be undertaken before the project can move from one phase to another. This also includes the final gate, which is the Financial Investment Decision (FID) when the project moves into the Execute phase. This is discussed in more detail throughout the subsequent chapters.

The project life-cycle is reviewed in this chapter and subsequent chapters to provide a clear observation of how the life-cycles alter as these LNG projects become more complex and, therefore, more difficult to benchmark. Rangnow (1994) points out that LNG developments, historically, have shown that there is limited sharing of benchmarking data across the industry.

2.4 Life-Cycles of Mega Complex LNG Projects

In a mega LNG (Liquefied Natural Gas) project, the project life-cycle is typically divided into multiple phases, each of which is critical for the successful execution and completion of the project. Combined with these phases, there are generally three main stage gates that help ensure the project progresses smoothly and meets its objectives. As demonstrated in Figure 2.1, these stage gates or 'risk gates' are at the end of each phase where the project goes through an intense scrutiny and governance process before it can pass through to the next phase. Cooper (2008) points out that the stage gates are a vital component of the governance or decision-making process; however, they may also lack teeth and projects are really killed at these gates.

As mentioned above, the most important stage gate is at the end of the Develop phase, and this is known as the FID for the project to transition into the Execute phase. Accordingly, Cooper (2008) suggests that, after the Assess (concept) phase, projects rarely get deferred, as they are then part of the overall company road map and the overall company financial plan.

2.5 Chapter Summary

This chapter has provided an example of what a project life-cycle involves, which will be further deliberated in Chapter 6 – Life-cycle Model.

It has also highlighted that the most important stage gate, which is at the end of the Develop phase, known as the FID. Consequently, the FID serves as the most critical evaluation point within the development of a project. This is where a decision is made for the project to proceed into the next phase, that of the Execute phase.

This chapter has also pointed out that there is or can be a pitfall in the management style when they want to accelerate the Develop phase to ascertain the FID as soon as possible. Rushing through critical decision-making processes for the FID approval at the end of the Develop phase often leads to inadequate understanding or maturity of the SoW, which may have a detrimental effect during the Execute phase and could cause schedule and cost overruns going forward.

Consequently, this chapter has also highlighted the multiple phases within the project life-cycle and the importance of the management of the stage gates to undertake the need for good governance, as these stage gates promote transparency, accountability, and risk management, ultimately contributing to the project's success and sustainability.

3 Schedule and Cost Overruns

3.0 Overview

Schedule and cost overruns are significant challenges in mega LNG projects, due to their complexity, economy of scale, and the numerous factors influencing their execution. This involves the construction of complex facilities for liquefying and exporting natural gas, often requiring significant investments and technical expertise. To mitigate these challenges, project stakeholders often employ risk management strategies, conduct thorough feasibility studies, engage in early stakeholder consultations, adopt agile project management methodologies, and foster collaboration among JVPS, project teams, contractors, and regulatory authorities. Despite the best efforts to manage schedule and cost overruns in mega LNG projects, these challenges often persist due to the inherent volatility and complexity of such large-scale undertakings. Addressing these challenges requires a combination of detailed planning, robust project management, effective stakeholder engagement, and strategic risk management. By proactively addressing these factors, it is possible to mitigate the impact of overruns and improve the chances of successful project outcomes.

Organisations continue to struggle with schedule and cost overruns on mega complex LNG projects although they have gone through stage gate risk assessments. Ashkanani and Franzoi (2022; p 141) state, "Efficient risk management is challenging for the success of megaprojects. It helps minimize delays and cost overruns, improve project quality, improve decision-making processes and enhance execution performance."

The phenomenon of major cost and schedule overruns becoming particularly visible during the Execute phase of oil and gas mega LNG projects is a common challenge. This phase often reveals the full extent of issues that were either underestimated or not fully anticipated in earlier stages. Several factors contribute to this visibility, many of which are detailed across various chapters of this book. Olaniran et al. (2015) point out that, despite the innovations in management practices, cost and schedule overruns persist as the major concern within the oil and gas industry developing mega projects. By delving into these factors throughout this book, consequently, stakeholders in oil and gas mega LNG projects can gain insights into the root causes of schedule and cost overruns and develop strategies to address them effectively during the execution phase.

DOI: 10.1201/9781003688334-3

3.1 Chapter Introduction

The visibility of cost and schedule overruns during the Execute phase of mega LNG projects is often linked to deficiencies in the definition of the SoW during the Develop phase. This lack of definition can significantly contribute to schedule and budget overruns in mega LNG projects. To mitigate these challenges, it is crucial for project stakeholders to prioritise robust scope definition and management during the Develop phase of mega LNG projects. This includes conducting thorough feasibility studies, engaging stakeholders early in the process, defining clear project objectives, deliverables, and establishing mechanisms for scope change management and communication throughout the project life-cycle.

These overruns can also be influenced by management wanting to move through the Develop phase too quickly, to obtain the FID and commence the Execute phase. This pitfall is often due to the insufficient understanding of the SOW definition/preparation, rushed decision-making, and incomplete planning. To mitigate these issues, it is essential to focus on thorough planning, realistic decision-making, robust contract management, effective stakeholder engagement, and ongoing monitoring and control. Ashkanani and Franzoi (2022; p 141) state, "Better approaches are required to enhance the decision-making selection for the procurement process according to the type, area, scope, and size of the megaproject, aiming to avoid a shortage of materials and resources, delays and cost overruns".

For clarity, an overrun refers to a situation where the project exceeds its initial budget or schedule. It means that the actual costs incurred, or the time taken to complete the project, surpass the original estimates or targets set during the project's planning phase. This deviation from the planned budget or schedule can have significant implications for project stakeholders, including financial impacts, delays in project delivery, and potential reputational damage. Identifying the root causes of overruns and implementing effective mitigation strategies are essential for managing project risks and ensuring successful project outcomes. Addressing issues related to scope definition, planning and scheduling, risk management, contractual and procurement processes, and operational challenges can significantly reduce the likelihood of overruns and improve project performance.

3.2 Schedule Overruns

Schedule overruns on mega LNG projects refer to situations where the actual time taken to complete the project exceeds the originally planned or estimated duration. These overruns can result in delays in the start of LNG production and delivery, impacting the project's financial performance and the energy market.

Schedule overruns in mega LNG projects are not uncommon and can result in significant delays and cost overruns. LNG projects are highly complex and involve multiple stages, each of which presents its own set of challenges. These stages include exploration, production, liquefaction, transportation, and regasification. Various factors can contribute to schedule overruns in LNG projects. Given the scale, complexity, and interdependencies of LNG projects, even minor disruptions or delays in one stage can have cascading effects throughout the project life-cycle, leading to significant schedule overruns. To address these challenges, project stakeholders must adopt robust risk management strategies, conduct thorough feasibility studies, engage in early stakeholder consultations, and implement agile project management methodologies to mitigate risks and ensure timely project delivery.

Mega complex LNG projects often incorporate advanced cutting-edge technologies and experience unexpected or unforeseen technical challenges during the construction (Execute) phase. Solving these challenges can often lead to schedule extensions, cost increases and impact the overall life-cycle.

Given the critical importance of technical innovation in LNG projects, it is essential for project stakeholders to anticipate and proactively manage potential technical challenges throughout the project life-cycle. This includes conducting thorough risk assessments, fostering collaboration with technology partners, and maintaining flexibility in project schedules to accommodate unforeseen technical issues. By addressing these technical challenges, it is crucial to enable the successful implementation of cutting-edge technologies in mega LNG projects. Focusing on these challenges, project managers/teams can minimise schedule extensions and enhance the likelihood of successful project outcomes, even when working with advanced and evolving technologies in the LNG sector.

Local communities plus stakeholders may have concerns about the environmental impact, safety, and other aspects of developing LNG projects. Protests, legal challenges, and negotiations with stakeholders can delay project progress. Consequently, ineffective project management, poor communication, and inadequate risk management can all contribute to schedule overruns. Ashkanani and Franzoi (2022; p 141) state, ". . . schedule overruns during the execution of the project due to frequent scope changes, repetition of works, and added expectations leading to unforeseen issues and risks, which makes the project vulnerable to missing its objectives". To mitigate schedule overruns in LNG projects, it is essential to conduct thorough feasibility studies, establish robust risk management plans, engage with stakeholders early in the project life-cycle, and maintain effective project management and communications throughout all phases. By implementing these strategies, project stakeholders can proactively manage schedule risks, improve project predictability, and increase the likelihood of delivering LNG projects on time and within budget.

Additionally, project developers should anticipate potential challenges and build flexibility into project schedules to account for unexpected delays. Ashkanani and Franzoi (2022; p 143) state, "The entire decision-making process should focus primarily on achieving high quality. It may directly or indirectly handle challenges, mitigate issues and failures, leverage success factors and adequately incorporate lessons learned into developing and implementing enhanced megaproject management systems".

Due to the significant capital investments that experience substantial implications for the energy market associated with mega LNG projects, schedule overruns can lead to considerable financial and operational consequences. Therefore, careful planning, diligent execution, and adaptive project management are essential to deliver such projects on time and within budget. Consequently, stakeholders developing mega LNG projects must focus on rigorous project management, effective risk assessment and management, and transparent communication with all involved parties. Implementing contingency plans and maintaining flexibility in project execution can also help manage the impacts of schedule overruns. Ashkanani and Franzoi (2022; p 129) state, "There are multiple gaps associated with issues, failures, successes and challenges in industrial megaprojects. Improvements are needed in distinct management areas and over the entire project lifetime".

Mitigating schedule overruns requires effective project management practices, proactive risk assessment, contingency planning, and experienced project teams which are essential. Employing experienced project managers, utilising advanced project management software, conducting thorough risk assessments, and maintaining open communication among stakeholders are vital to identifying potential issues early and implementing appropriate mitigation measures. Ashkanani and Franzoi (2022; p 130) state, "Despite the importance of addressing issues and failures, there are also advantages in understanding the main factors that lead to successful megaprojects. This helps identify which changes are required and how they can be employed".

3.3 Cost Overruns

Organisations spend billions of dollars on schedule and cost overruns during the development of these complex LNG projects throughout their project life-cycle. To clarify, the meaning of 'project life-cycle' can be described as the logical sequence of major activities that a project goes through to accomplish its objectives. Cost overruns in LNG projects are not uncommon and can be attributed to various factors. LNG projects are typically large-scale and complex, involving multiple stages such as exploration, production, liquefaction, transportation, and regasification.

The investigation also discloses the underlying factors that could be triggered by using obsolete oil and gas benchmarking and management techniques, consequently causing cost overruns in mega/giga complex LNG projects. Additionally, effective stakeholder communication and risk mitigation strategies can help address potential challenges as they arise, reducing the likelihood of significant cost overruns.

Projects are sponsored using **CAP**ital **EX**penditure (CAPEX), which refers to the funds a company or organisation invests in acquiring, upgrading, or maintaining long-term assets that are essential for its operations. These assets are typically expected to provide economic benefits to the company over an extended period. CAPEX is also known as capital investment or capital spending. CAPEX is distinct from **OP**erating **EX**penditure (OPEX), which refers to the day-to-day expenses incurred to maintain and run a business, such as salaries, rent, utilities, and supplies.

Monitoring and managing CAPEX is crucial for businesses to ensure their investments are aligned with their long-term strategies and to optimise the utilisation of resources. It also plays a significant role in financial planning, budgeting, and decision-making processes within an organisation. Love et al. (2013) suggest that cost overruns should be separate from cost escalations, which is the anticipated growth of a budget due to things such as inflation, etc.

3.4 Chapter Summary

This chapter has reviewed schedule and cost overruns that occur when developing mega complex LNG projects throughout their life-cycle. It also explains why these schedule and cost overruns only become visible during the Execute phase of the project life-cycle. It also suggested that project teams and management should not put unrealistic expectations to achieve the FID and move into the Execute phase. By doing so, it may be conceivable that the full SoW is not fully understood or visible at this stage by fast tracking the Develop phase to achieve FID. Consequently, by rushing the Develop phase, this can contribute to these schedule and cost overruns in the Execute phase. More deliberation at the stage gate at the end of the Develop phase needs to be part of the governance process, both project and corporate, to scrutinise whether the project is really ready to commence the move into the FID. If there are still risks in the development phase, these need to become visible in the project risk register and fully scoped out as to what these risks are and the consequence of moving these risks into the e phase. This may need a mind-set change to have the team to fully ascertain the schedule and cost risks and the implications of this to become part of the execute budget, schedule and subsequently the impact on the RFSU. It is also essential that the governance team (corporate and project) check to ensure the schedule logic is intact, i.e., all activities are fully logically linked, therefore capturing the risks that get moved over.

4 Benchmarking in the LNG Sector

4.0 Overview

Benchmarking has evolved over the last few decades in the oil and gas (LNG) industry as a more acceptable strategic tool for organisations to seek continuous improvement and measurement against other organisations, which strive to be, or are already, the best-in-class. Horváthová et al. (2021; p 1) state, "Benchmarking is part of the concept of quality management. It is a method of analysing and comparing practices and experiences in various areas of business operations." Prior to this, Singh et al. (2017; p S617) state, "The essence of benchmarking is the process of identifying the highest standards of superiority for products, services or processes and then make the essential improvements to reach those standards."

According to Macneil et al. (1995), benchmarking is well-defined as a method of continuous improvement that should involve systematic evaluations of the processes to represent best practices. Therefore, what is best practice benchmarking for mega and giga LNG projects? Remington and Pollack (2012; p 32) state, ". . . contributor to project failure has been identified as poor risk analysis. This applies to planning and implementation phases." This is defined as the benchmarking process that is transparent and specific to the project's size, location, infrastructure, and economy of scale. According to Anand and Kodali (2008), benchmarking as a tool has more practical or industrial utility than academic/research utility. Furthermore, benchmarking is a valuable practice in the LNG sector that involves comparing the performance, processes, and key performance indicators (KPIs) of one LNG project or facility to those of similar projects or industry standards.

4.1 Chapter Introduction

The objective of benchmarking in the LNG sector is to determine the best practices, areas for improvement, and opportunities to enhance efficiency, reduce costs, and increase competitiveness. In addition, benchmarking in the LNG sector is a strategic tool that allows companies to assess their performance, identify opportunities for improvement, and stay competitive in a dynamic and evolving industry. It requires access to relevant data, industry standards, and a commitment to

DOI: 10.1201/9781003688334-4

continuous improvement to provide value to LNG projects and facilities. By incorporating relevant data, adherence to industry standards, and a dedication to continuous improvement, stakeholders can effectively deliver value in LNG projects and facilities. This approach encourages innovation, sustainability, and long-term success in the dynamic LNG sector.

4.2 Benchmarking History

Benchmarking has a long history, used by surveyors, stone masons, fishermen, and tailors as a method of measuring from a known point to another, thus the term benchmark. According to Andersen and Pettersen (1996), there are many theories about benchmarking and where the process really came from; one of the claims is that benchmarking originated from geographical surveying, where the benchmark was a topological reference point in the terrain. Wherever, or however, the benchmarking process originated, the present-day use of benchmarking in the oil and gas industry is becoming an essential component of the project management process, particularly when benchmarking mega complex LNG projects. Andersen and Pettersen (1996) suggest the original meaning of the word benchmark is described as a predefined position to start from as a reference point to measure against. It is, however, essential for organisations to be able to measure their projects for cost and schedule against similar projects, those already completed or nearing completion, and to recognise the types of benchmarking requirements. If benchmarking is correctly promoted within an organisation, with genuine commitment and ownership from senior management to ensure the process is clearly established, it motivates change and improvements within that organisation.

There are commonly four distinct types of benchmarking:

- internal (which means measuring and comparing company data);
- external (which can be considered competitive);
- functional (which can be considered non-competitive); and
- standard (which is measuring against best-in-class or generic-type benchmarking).

According to Oakland (2007), benchmarking has been described as a reference or measurement standard used for comparison and is a continuous process of measuring against best practices and processes. Over the past few decades, benchmarking has been known as the process of continuous improvement and a structured means to allow organisations to develop performance measurement systems which allows them to produce the best-in-class projects. Macneil et al. (1995) describe the process of benchmarking as a method for continuous improvement, and it is a formal method of striving to become best in class. Benchmarking is a fundamental tool for businesses seeking to remain competitive and improve their performance. It continues to evolve as industries change and new technologies emerge, helping organisations adapt to the ever-changing business landscape.

4.3 Benchmarking Framework

A benchmarking framework is a structured approach that organisations use to compare their performance, processes, or practices with those of other organisations or industry standards. This framework provides a systematic method for gathering data, analysing results, and identifying areas for improvement. By following this systematic technique, stakeholders can effectively enhance operational efficiency, ensure compliance with industry standards, and deliver sustainable value in LNG projects and facilities.

The benchmarking framework also provides a structured approach to improving performance and processes by utilising the best practices of other organisations. Depending on the specific objectives and areas of focus, organisations may modify and adapt this framework to suit their unique needs and challenges. The benchmarking framework builds on the development of the life-cycle model of phase performance measurement. Kulikowski (2020; p 724) states ". . . the standard benchmarked model is reductive in the sense that it does not represent all details or nuances of benchmarking procedures but thanks to its universality, it might be applied in a broad organisational and business context and thus constitute a good starting point to expand it into the evidence-based benchmarking model." As such, it provides a benchmarking proposal for future mega/giga complex LNG projects to allow organisations to understand what is required to make these multifaceted projects a success using quantifiable benchmarking that is developed specifically for mega complex LNG projects. According to Marrewijk et al. (2008), the Project Management Book of Knowledge (PMBOK) guide's integrative perspective is too limited to understand the dynamics of the project culture in mega projects. The conceptual frameworks provided suggest that it is essential for organisations to apply a structured governance role early in the project life-cycle, which has a strong link between corporate and project governance. Cooper (2008) suggests that defining governance roles and responsibilities is an important facet of stage-gates.

The author's complexity framework in Figure 4.1 identifies the complexity between the following definitions of projects: small conventional projects, large conventional projects, mega projects, and giga projects (complexity versus size). This framework aims to enable organisations to determine how their project is defined, based on the size of the project, and, as such, determine if they are mega or giga projects due to their complexity ratio.

The typical stages of the benchmarking structure are displayed in Figure 4.2. According to Oakland (2007), this benchmarking framework has five main stages (Plan, Collect, Analyse, Adapt, and Review), which are all focused on trying to measure comparisons and identify areas of action and change.

Oakland (2007; pp 154–155) suggests that some of the major portions of the five stages are as follows:

Plan the Study:

- select processes for benchmarking, bring together the appropriate team, identify benchmarks for data collection, and identify best competitors or operators of the process.

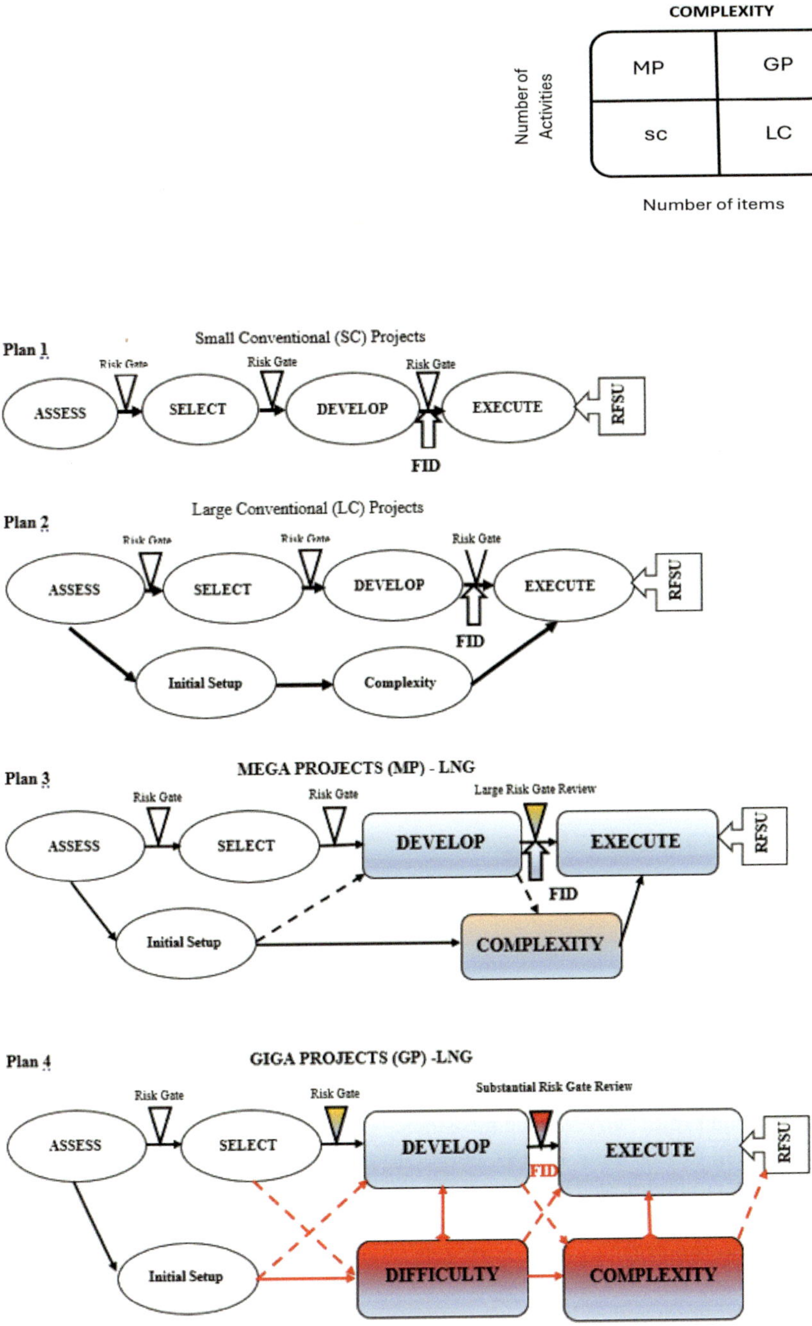

Figure 4.1 Project Complexity Framework

Source: Author's Research

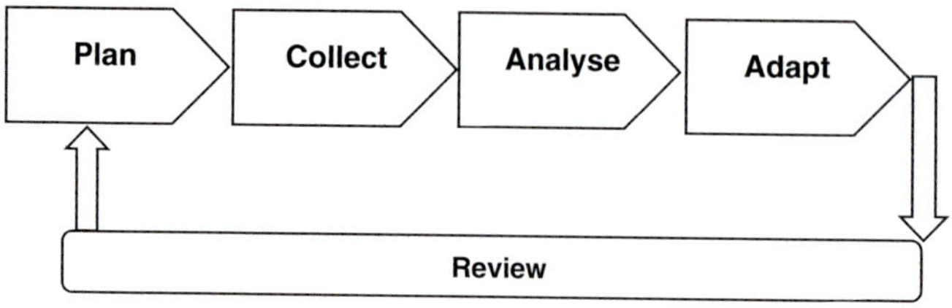

Figure 4.2 Five Main Stages of Benchmarking
Oakland (2007; p 154)

Collection of Data and Information:

- decide information and data collection methodology, record performance levels, identify benchmarking partners, conduct primary investigations, use data collection methodology, and conduct site visits and discussions.

Analyse:

- normalise the performance data, develop a comparison matrix, isolate, and understand enablers.

Adapt:

- catalogue information, develop performance levels, incorporate best practice enablers, identify barriers of change, develop action plans, and implement and integrate actions within the organisation.

Review:

- assess outcomes, review benchmarks, and share experiences.

Building on Oakland's (2007) framework, it is proposed that, when benchmarking these multidimensional LNG projects, there is a need to provide tighter control and alignment of the process for each stage within the framework by introducing corporate and project governance measures.

According to Flyvbjerg (2014), corporate governance is becoming a means of enforcing accountability. The proposed framework depicted in Figure 4.3 includes in-depth project governance checks at each stage, with an overall compliance alignment to the organisation's corporate strategy. Merrow (2012) suggests that good governance requires decision-makers to be aligned with the long-term health of the organisation and be responsible for securing the correct resources to enable the projects to be developed safely and correctly.

Priemus et al. (2008) suggest that decision-making follows analysis; there is no decision-making without the right information and analysis.

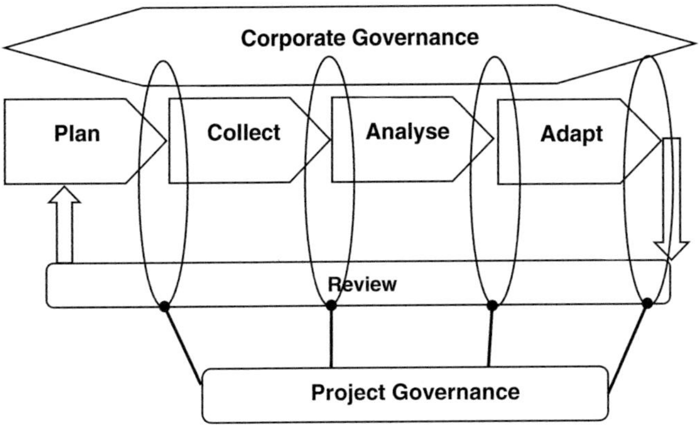

Figure 4.3 Proposed Framework for Mega or Giga LNG Projects

Author's Research

These two supplementary requirements, i.e., corporate and project governance, illustrated in Figure 4.3, provide the alignment between the project and the organisation's strategic values. Flyvbjerg (2014) implies that, although good governance processes are slow, the trend of good governance is gaining a foothold even in megaproject management.

By enhancing this process, it allows the project manager and their team to reassess the project governance team on how they have arrived at their conclusions at the end of each stage. Priemus et al. (2008) propose that governance means setting up a structure which contains a set of decision-making processes and methods for accumulating knowledge to ensure that creativity and discipline are brought to bear.

Organisations' corporate and project governance processes and alignments were examined to determine how they were implemented or aligned with each other. By conducting thorough reviews of corporate and project governance processes and aligning them with each other, organisations can enhance their overall management practices in LNG projects and facilities. This alignment enhances decision-making, promotes accountability, and supports the implementation of strategic objectives while ensuring compliance and effective risk management. Mbassegue et al. (2017; p 11) state, "There is also a need for governance, which includes integrating the various stakeholders at all stages of the megaproject." Consequently, it is suggested that organisations undertaking mega complex LNG projects should adopt a more transparent or practical approach regarding the project and corporate governance. Kerzner (2014) suggests that organisation executives understand the business language but do not listen to project management techno talk. Therefore, aligning the corporate and project governance management teams requires good pragmatic communication between both entities. Mbassegue et al. (2017; p 12) state, "The complexity of megaprojects induces an understanding of management practices and modalities based on the diversity of case studies." It is essential that the organisation's corporate values regarding governance be part of the project. Sanderson (2012) suggests that governance solutions require a consciously designed project culture to encourage trusting and collaborative behaviour and, with formal governance, to elaborate further, a representative of the organisation's corporate governance team should be embedded within the project team. Such a move may be viewed by the project director as a slight on their ability to maintain the alignment between corporate and project needs and may engender an atmosphere of distrust. Sanderson (2012) suggests that focus on project governing should not replace the focus on project governance; nonetheless, it should be for the relevance to the project. However, if the context of this positioning is open and honest, the project director and the teams benefit from the knowledge and assistance that aligns the corporate and project governance needs.

Furthermore, project and corporate governance are the means of allowing the organisation's decision-makers to ensure that the project remains aligned with the organisation's corporate values. Priemus et al. (2008) propose that governance processes set up decision-making frameworks to make sure all the right questions are being asked within the project. This process is remarkably like the stage gates at the end of each phase of the project life-cycle. Refer to Figure 3.3 for more details within each individual phase of the project life-cycle.

Project governance is implemented at the end of each stage and during the review period within each phase of the project life-cycle. By applying project governance consistently at the end of each project stage and throughout the review periods within each phase of the project life-cycle, organisations can enhance project success, mitigate risks, and achieve desired outcomes efficiently and effectively. It provides a structured approach to project management that enables successful project delivery. According to Kerzner (2013), project governance focuses on how to keep the project on track. Kerzner (2013) goes on to explain that the project governance is more operationally directed with decisions based upon the predefined parameters on project scope, time, cost, and functionality.

Corporate governance overarches the five stages of the framework, which ensures the project aligns with the organisation's strategic compliance process within each phase of the project life-cycle. It provides the framework for effective oversight, risk management, and decision-making, ultimately contributing to the success and sustainability of the organisation's initiatives. Kerzner (2013) implies that corporate governance focuses on how well an organisation's portfolio is aligned and how it satisfies the overall business objectives. As suggested by Kerzner (2013), corporate governance provides strategic direction with a focus on how project success satisfies the corporate objectives. Adapting Oakland's (2007) data of the five-stage model for mega LNG projects, the major components of the five stages are as shown in Figure 4.3, as well as incorporating the following additional processes.

Plan the Project: during the planning phase of any project, a project governance review at the end of the stage and corporate governance alignment is essential. Accordingly, this is the basis of the project where each phase is benchmarked against previous projects. The benchmarking data is used to develop the phase durations and build up the entire project life-cycle to the Ready For Start Up (RFSU). The schedule alignment is critical to the project's success, which requires the master schedule to be baselined ensuring correct benchmarking data is used that is appropriate to the mega complex project being developed.

Collect data and information: this involves gathering relevant project data from reputable benchmarking organisations, such as Independent Project Analysis (IPA) and the Performance Forum (PF), According to Nalewaik and Mills (2015), the correct benchmarking data may be difficult to find due to the complexity of these multifaceted LNG projects. This may also be associated with the lack of sharing of benchmarking data linked to these complex projects.

Analyse the information and data:

- project governance review at the end of the stage; and
- corporate governance alignment.

Ensure that the facts being entered into a matrix for comparisons against other benchmarking projects is comparing appropriate data or projects for the current mega complex project being developed. Nalewaik and Mills (2015) propose that, when analysing data, it is important to use an independent third-party agent to ensure the correct, unbiased data is being used.

Adapt the approach going forward:

- project governance review at the end of the stage; and
- corporate governance alignment.

This is accomplished by cataloguing the information, allowing the project team/manager to create a benchmarking profile that can be used on the project. One concern is the lack of current benchmarking data, which is deemed dependable enough to be used to benchmark mega complex LNG projects. Nalewaik and Mills (2015) indicate that when conducting benchmarking, the data needs to be relevant to the project objectives or size, i.e., mega complex-sized projects, not conventional projects.

Review the status of the project:

- project governance review at the end of the stage; and
- corporate governance alignment.

As identified in Figure 4.4, the governance process needs to be repeated throughout all phases of the project life-cycle to ensure that the project governance process complies with the organisation's overall strategic requirements. According to Sainati et al. (2017), governance is a complex multi-level concept that plays a key role in the management of mega projects.

For mega complex LNG projects, it is important that this process is strictly managed to allow the decision-makers the opportunity to review, challenge, agree, or

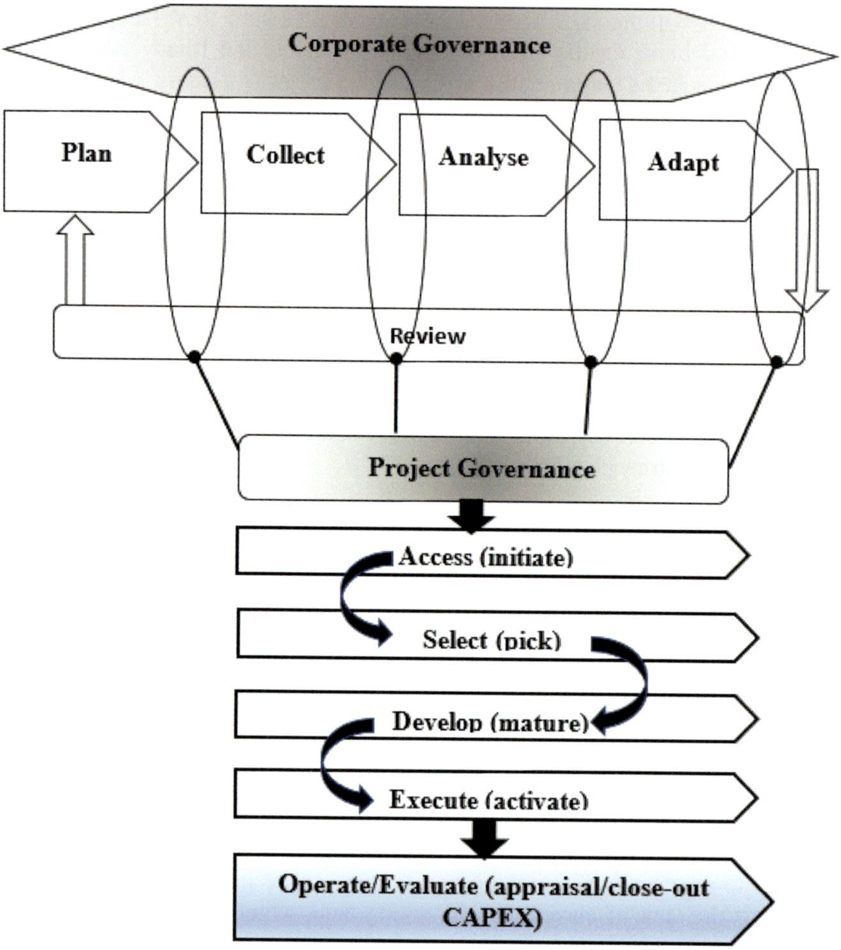

Figure 4.4 Process Repeated Throughout the Project Life-Cycle.

Author's Research

disagree on the proposed outcomes of the project process being put forward. This approach helps mitigate risks, ensures alignment with strategic goals, promotes transparency, and enhances overall project success and sustainability. Priemus et al. (2008) recognise that a fundamental aspect of the governance regime is that at least three alternative concepts need to be deliberated at the initial stages when options are still open.

Accordingly, during the Assess (initiate) phase and the development of the Project Management Plan (PMP), the overall life-cycle duration to RFSU is benchmarked at this early phase, as shown in Figure 4.5. Overall, benchmarking the life-cycle duration to RFSU during the Assess phase and the development of the

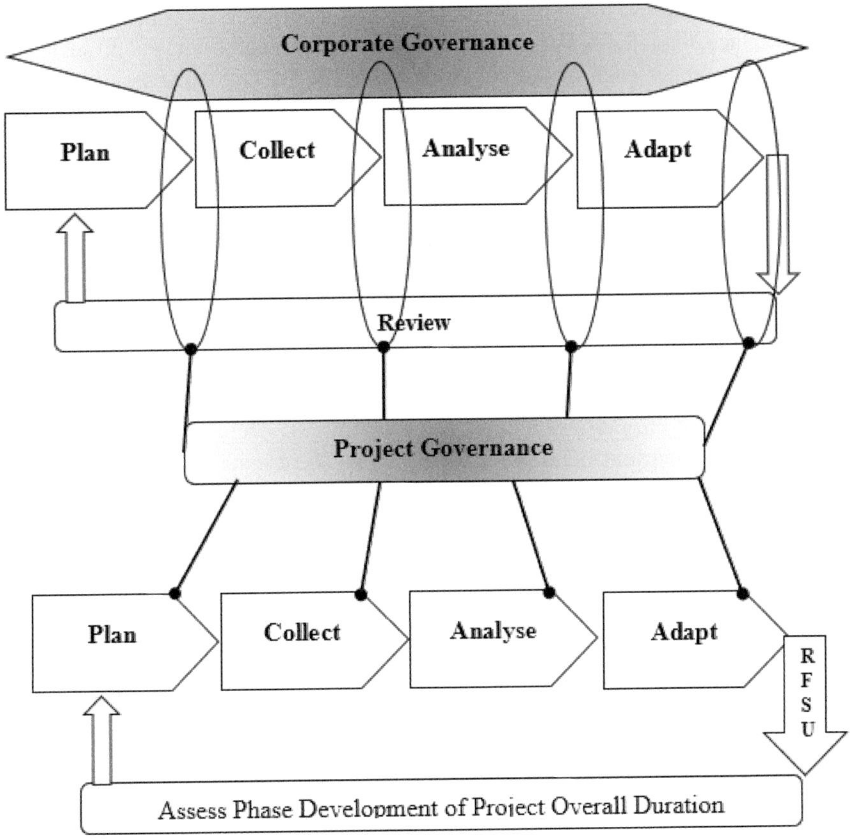

Figure 4.5 Project Duration – Level 1 Development to RFSU

Author's Research

Project Management Plan is fundamental to laying a solid foundation for successful project execution. It facilitates informed decision-making, enhances planning and control efforts, and supports effective communication with stakeholders throughout the project life-cycle. In essence, benchmarking the life-cycle durations to RFSU facilitates the development of comprehensive cost estimates at level one and level two. These estimates serve as critical inputs for planning, budgeting, risk management, and decision-making, ultimately contributing to the successful execution and completion of the project within expected cost parameters. Consequently, this allows the decision-makers the opportunity to evaluate the overall project schedule and cost estimates of the overall project proposal. Priemus et al. (2008) suggest that governance processes for mega projects are spread over a lengthy period of time; therefore, there is an opportunity to transform the governance structure as the project unfolds.

This high-level overview of the proposed mega complex LNG project governance interface, depicted in Figure 4.5, incorporates both corporate and project governance throughout each of the stages and into the life-cycle phases. According to Kerzner (2014), governance relates to those decisions that define expectations, accountability, responsibility, or verifying performance. This then provides the platform for effective decision-making within the project to be undertaken. Establishing a clear platform for decision-making is crucial for the success of any project. When everyone involved understands the framework and criteria for making decisions, it streamlines the process and ensures that decisions are made efficiently and effectively. This clarity helps in aligning goals, managing risks, and fostering collaboration among team members. Ultimately, it contributes to the overall success of the project by promoting informed and timely decision-making.

According to Merrow (2003), good governance is about the accountability of management for the performance of the corporation. Arena and Molloy (2010) propose that the current literature available is on the dynamics of the early planning phases of mega projects and, as such, does not provide details of the different durations to complete the actual individual phases that make up the mega project life-cycle.

Consequently, as the project progresses through the phases, such as planning, executing, monitoring, and closing, it is essential to adjust the benchmarking process accordingly. This means understanding that each phase may have different durations and costs associated with it.

For example, the Execute (execution) phase often tends to be the longest and most resource-intensive phase in many projects. During this phase, benchmarking would focus on comparing actual progress and costs against planned milestones and budgets. This helps in identifying any deviations or areas where adjustments may be needed to keep the project on track.

Incorporating extended durations into benchmarking means acknowledging that certain phases may take longer than others, and therefore, benchmarks for time, cost, quality, and other parameters should be realistic and aligned with the specific requirements of each phase. This ensures that benchmarks remain relevant and useful throughout the entire project life-cycle, supporting effective decision-making and overall project success.

Plan the phase: this is the part of the project where each phase is benchmarked against previous projects. The benchmarking data is used to develop the phase durations and build up the entire project life-cycle to the Ready For Start Up (RFSU). Merrow (2012) indicates that at the Assess phase (gate 1) review, the cost estimate and schedule forecasts do not mean much. However, it is at this stage that the Work Breakdown Structure (WBS) development is linked to the Cost Breakdown Structure (CBS) to form the basis of a roadmap for further development.

Within this planning phase, the project team develops a strategy and collection methodology which is relevant to the project being undertaken.

Schedule alignment: this process is critical to the project's success and is, therefore, an essential requirement to ensure that the schedule is only baselined once, and the agreed benchmarking data incorporated within the plan. This, however, does not mean that the project cannot be re-baselined as it becomes more mature, should the need arise.

When this framework is applied to an LNG mega complex project, it is essential that benchmarking is realistic. It is also essential that:

- benchmarking must be an on-going process throughout each part or phase of the project life-cycle methodology. Hence, it should be revisited at each phase gate review, before moving into the next phase of the project life-cycle; and
- benchmarking does not encapsulate the 'tick-box' culture. It is part of a full due diligence process of risk management.

This also identifies whether a project is robust enough on cost, schedule, contracting strategy, and, most importantly, the organisation structure required to move through a stage-gate (decision point/assurance risk) into the next phase. Drennan et al. (2014) suggest that risk management is not to be a tick in the box, but it is a mechanism to provide assurance that risks are being managed.

Data and Information Collection: this comprises gathering relevant project data from benchmarking organisations, such as Independent Project Analysis (IPA) and the Performance Forum (PF), who are independent and sanitise the data before engaging with organisations. Relevant benchmarking data is identified, which can be used for the project being undertaken. This builds on the Data/Information strategy and methodology, which is proposed in the planning stage.

Koskinen et al. (2003) suggest that a project has a combination of tacit knowledge and benchmarking, where the members of a project team can interact face-to-face with each other to understand and reinforce tacit knowledge sharing. Data and information collection should also include tacit information gathering, which is the knowledge and experience those individuals bring to an organisation from previous projects, etc.

Analysing the Data: is when the data is used in a matrix for comparisons against other benchmarking information, at which stage it is determined if it is appropriate for the current mega project development.

Adapt the Data: by cataloguing the information which allows the project to create a relative benchmarking profile to be used on the project.

Data review: as shown in Figures 4.2–4.5, the review of data is an ongoing process throughout the phases. This review enables the project team to understand the information being utilised for the project, how this data, provided by benchmarking companies, is used, and agreeing on how relevant the data is to the current project. Tacit information is compared to the quantifiable data received from the independent benchmarking organisations. Koskinen et al. (2003) imply that tacit knowledge is practical. Tacit knowledge comes from experience which is deeply rooted within an individual's mind and consists of schematic beliefs or perceptions from individuals. Koskinen et al. (2003) go on to suggest that this tacit knowledge forms part of a person's competency to conduct the work.

Project Governance review: this is one of the most important processes of this proposed framework. It is where the Project Manager and team discuss openly with the project governance team how they arrived at the benchmarking decision and why they have selected the benchmarking data to be used on their project. Koskinen et al. (2003) suggest that seldom does one person know enough to solve complex problems faced in projects; therefore, team members rely on each other for knowledge and advice.

Corporate Governance: it is essential that the project governance process is aligned with the corporate governance process. This ensures that the organisation's goals and strategic values are incorporated into all phases of the project and the overall project duration. According to Sanderson (2012), mega projects perform badly and underperform in the presence of incoherent or underdeveloped governance arrangements.

4.4 Analysis of Benchmarking Mega Complex LNG Projects

Benchmarking of mega complex LNG projects within Australia is difficult to manage due to the remote locations of the gas fields, which cause the LNG supply chain to become more complex and more expensive to develop. According to Söderlund et al. (2017), little attention has been focused on what happens within mega projects, such as the challenges of starting a mega project, how managers work to establish functioning regimes for cooperation, including the relevant mechanisms for coordination, and the ongoing practice of managers coping with complexity challenges. These complexities can be characterised as environmental requirements, the economy of scale, and site remoteness. This brings about the potential for pitfalls which can occur during the benchmarking process, such as lack of focus, direction, and understanding of what is being measured and who has the overall ownership of the process within both management and the project team members. While onshore process plants may differ in terms of their core functions or operations, the supporting factors can vary significantly. These factors encompass a wide range of factors that can influence the design, operation, and overall effectiveness of the plant. However, as Stapenhurst (2009) implied, no two projects are ever the same, but aspects of all projects can be benchmarked using normalisation techniques. Such normalisation techniques should include removing inclement weather, labour disputes, and machinery/equipment downtime to enable comparison of raw data. However, in many projects, benchmarking fails due to the lack of commitment or ownership of the process by senior management. Without their commitment, it is difficult to allocate resources, set priorities, or drive the necessary changes according to benchmarking results.

Senior management plays a crucial role in ensuring that benchmarking efforts are aligned with strategic goals and that the insights gained are effectively used to improve performance. When they don't prioritise benchmarking, it can lead to lack of resources, inconsistent implementation, or even ignoring the findings altogether.

Benchmarking mega complex LNG projects still needs to be understood as to where this benchmarking data comes from, for instance, is it from smaller LNG projects with tacit information being applied, as most, if not all, the mega complex LNG projects have been unable to meet their schedule and budget requirements.

Managers need to discuss with their teams how they intend to work with the benchmarking data and performance measurement systems and how relevant it is to their organisation/project. This also brings about the misunderstanding within some organisations that have benchmarked at the start (Initiate phase) of the project that believe there is no need to revisit the benchmarking throughout the other phases as the project matures going forward. While benchmarking at the beginning of a project provides a necessary baseline, it is essential to revisit and update benchmarks periodically throughout the project life-cycle. This practice enables organisations to remain agile, responsive to changes, and continuously improve their project performance.

4.4.1 Types of Benchmarking of LNG Projects

When it comes to the phases of mega complex LNG project life-cycles, there is little benchmarking data available for the individual phase durations of these complex LNG projects, besides tacit information. Consequently, the literature pertaining to these complex LNG onshore projects has always been sparse, especially regarding the life-cycle benchmarking of these onshore plants for schedule, cost, and quality. While benchmarking individual phase durations in mega LNG projects may present challenges due to complexity and confidentiality concerns, organisations can leverage industry collaboration, research, and internal continuous improvement efforts to enhance their understanding and optimise project execution. Remington and Pollack (2012; p 3) state, "The management literature has, in recent years, made considerable theoretical advances in our understanding of complexity, but this hasn't until now been applied to complex projects – or at least, not in a way that might help us actually manage those projects separately."

The number of LNG projects in the oil and gas industry has increased over the past decade and will continue to do so for the near future. Consequently, this highlights the challenge of delivering these projects on time and budget. According to Nooteboom (2004), the oil and gas industry continues to grapple with how to bring a balance to the planning and execution of these complex developments in terms of effective contracting, coordination, risk allocation, and conflict resolution.

The use of project management techniques and processes in the construction industry has been in place for centuries, becoming a more prevalent and visible process over the past few decades in many industries, particularly in oil and gas (LNG) projects. While project management techniques have been integral to the construction industry for centuries, their application has expanded and intensified in modern times, particularly in complex sectors like oil and gas, including LNG projects. This evolution underscores the importance of structured, systematic approaches to managing projects to achieve success amidst growing complexity and global competitiveness.

Currently, the use of benchmarking data still forms a critical part of Total Quality Management (TQM), and the project management processes, allowing organisations to compare their projects against other similar projects. Zairi (2011) proposes that effective benchmarking equips the project team and senior

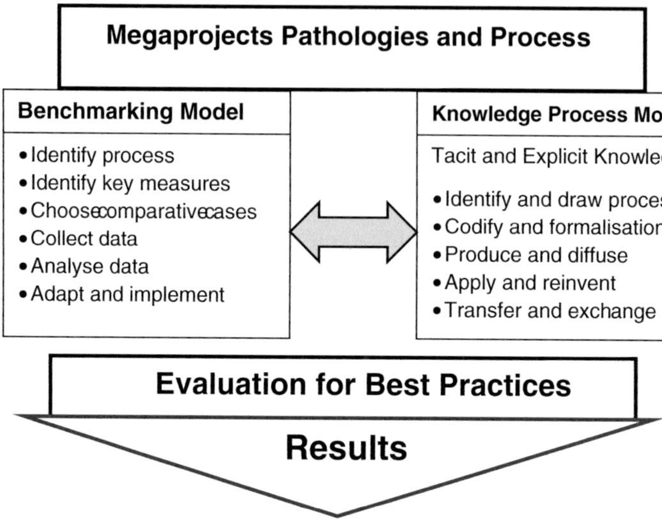

Figure 4.6 Model of Knowledge – Benchmarking for Mega Projects

Mbassegue et al. (2017; p 13).

management with the necessary capability to enable them to manage their projects effectively. As the techniques over the decades have been refined, benchmarking of most of these complex projects is against their phases or stages throughout their life-cycle. Mbassegue et al. (2017; p 13) state, ". . . by implementing a benchmarking approach based on knowledge management on processes implemented in megaprojects, the teams involved will be able to improve from project to project."

The knowledge management benchmarking process can be applied to mega projects as described in Figure 4.6.

4.4.2 *Structure of LNG Benchmarking*

A more specific or pragmatic approach is required when ascertaining benchmarking data for time, cost, and quality of each phase of the project life-cycle for mega LNG projects due to their complexity. By adopting this specific and pragmatic approach, organisations can effectively ascertain benchmarking data for time, cost, and quality at each phase of mega/giga LNG projects, thereby enhancing project management and overall success. This would include the use of tacit information, which would be slightly more prominent when determining a benchmarking methodology for the life-cycle phases of projects. By integrating tacit information into the benchmarking process, project managers can develop more robust and realistic benchmarks for time, cost, and quality across the life-cycle phases of mega/giga LNG projects. This approach not only enhances project performance but also mitigates risks associated with project complexity. Westland (2007) suggests that the most critical phase of a project life-cycle is the initiation (Assess) phase – if not rigorously evaluated, the risk of project failure can be greater. Consequently, the Assess phase is the first step of the development of the project. This is where the management team considers options/concepts to move through the risk gates or stage gates into the next phase (Select Phase).

The completion of the Select phase is when the crucial decision-making occurs in order to choose the correct option to move through the second risk gate into the Develop phase. This refers to the decision-making process that occurs at the second risk gate. It is critical because it determines whether the project is ready to move from planning into the detailed development stages of the Basis of Design (BoD) and Front-End Engineering Design (FEED). This typically involves establishing the fundamental requirements and specifications for the project. It sets the foundation for further detailed design and development, and then the FEED, this phase follows the BoD and involves detailed engineering design work. It is where the project begins to take shape in a robust manner. These are developed for the risk review, which is required to be undertaken to reach the Financial Investment Decision (FID) approval. Rolstadas et al. (2011) point out that, prior to the FID decision, it is important to invest time and effort to develop a robust project definition to improve the predictability of the overall scope before the approval of FID.

4.4.3 Benchmarking and Large and Small LNG Projects

Benchmarking is becoming increasingly important to the oil and gas industry, particularly in the development of mega complex LNG projects. Overall, benchmarking in the context of mega LNG projects enables companies to learn from past experiences, improve project outcomes, and navigate challenges effectively in a competitive global market. It serves as a strategic tool for continuous improvement and sustainable development in the oil and gas sector. Victor et al. (2006) suggest that a minimum economic scale for new LNG plants is two or three trains with around 7.5 million tonnes per annum (mtpa) total capacities. As with the economy of scale, which increases the size/cost/schedule and complexity of these projects, it is critical to ascertain that the benchmarking is accurate by using quantifiable data. Leveraging quantifiable data for benchmarking in mega LNG projects ensures that comparisons are objective, trends are identified accurately, biases are minimised, decisions are well informed, continuous improvement is facilitated, and credibility is maintained. This approach is indispensable for navigating the complexities and scale of mega complex LNG projects effectively. The financial dynamics of mega complex LNG projects require stakeholders to adopt a strategic and patient approach, managing risks diligently while focusing on achieving operational milestones that pave the way for revenue generation and eventual financial returns post the Operate phase. With this large investment outlay, stakeholders, or Joint Venture Partners (JVPs), need to be confident in the accuracy of the benchmarking data being prepared for the specific project to achieve a successful outcome on time and budget. Ultimately, precise benchmarking data instils confidence among stakeholders and JVPs by providing a robust foundation for strategic planning, risk management, decision-making, and performance evaluation in mega LNG projects. It supports the achievement of project goals and enhances the likelihood of delivering successful outcomes on time and within budget constraints. Invernizzi et al. (2018) point out that mega projects share characteristics of being extremely expensive and long, including being politically sensitive, and involve many internal/external stakeholders. Consequently, detailed benchmarking data that reflects the durations of each phase of the project life-cycle enables JVPs to make informed decisions, manage risks effectively, allocate resources efficiently, negotiate contracts confidently, monitor performance rigorously, drive continuous improvement, and uphold transparency and accountability throughout the development of mega LNG projects. Invernizzi et al. (2018) suggest that benchmarking offers significant potential to investigate the characteristics that have the most impact on the project performance.

Whether the LNG project is called a mega or giga complex project, depends on the reservoir size and the reserves available which is determined at the drilling of the development wells. Consequently, the classification of an LNG project depends on a combination of factors, including reservoir size, reserve estimates, technical complexities, and economic considerations. The drilling and evaluation of development wells play a pivotal role in determining these factors and ultimately shaping the scale and classification of the LNG project. These activities require careful planning, advanced technology, and expertise to optimise outcomes and mitigate risks effectively. It is considered that the current mega LNG project would produce (per train) up to 3 mtpa (million tonnes per annum). Mokhatab et al. (2006) point out that small-scale LNG plants have capacities as low as 0.01 mtpa and are still economically viable. Subsequently, it is therefore projected that a future giga LNG project would produce between 4–8 mtpa (per train) or greater. A giga LNG project represents a significant milestone in the LNG industry, characterised by its immense production capacity per train and overall economy of scale. These projects are designed to meet growing global demand for LNG while addressing complex technical, environmental, and economic considerations.

As most mega/giga complex LNG projects continuously overrun on cost and schedule, more emphasis needs to be focused on accurate benchmarking for individual LNG projects. Prioritising accurate benchmarking for mega/giga LNG projects is critical for managing costs, adhering to schedules, mitigating risks, supporting decision-making, driving continuous improvement, and building stakeholder confidence. It enables project teams to navigate the complexities and challenges inherent in large-scale LNG developments effectively, ultimately improving project outcomes and maximising long-term value. Although benchmarking has become, and should remain, one of the most important processes within the TQM suite of tools, it needs to be a more prominent feature of the risk assessments for future projects. Olaniran et al. (2015) point out that despite the innovations in management practices, cost and schedule overruns persist as the major concern within the oil and gas industry developing mega-projects.

One of the roles of project managers is to ascertain if accurate benchmarking is undertaken during the planning of each of the project phases (phase-based benchmarking) and to determine the importance to the success of a project, especially in mega/giga complex LNG projects. Olaniran et al. (2015) point out that problems created during the planning could be a calamitous issue during the construction of the project (execute phase). The fundamental issues could stem from the fact that benchmarking is an expensive process that requires the commitment of funds and guaranteed sponsorship of senior management and their stakeholders. For this reason, benchmarking consumes a lot of resources (time and costs) and is an expensive process to establish and maintain. Therefore, most organisations tend to shortcut the process by using more tacit knowledge to force the result and hold the end date for the RFSU. However, tacit information has not been sanitised which can result in misinformation being utilised on a project rather than quantifiable benchmarking data to allow the comparison of like for like. Validating information before using it in a project, reduces the risk of misinformation and ensure that decisions are based on reliable data rather than assumptions or incomplete knowledge. This approach improves the overall integrity and effectiveness of the project outcomes. Cortada and Woods (2000) suggest that tacit knowledge, or expertise, from individuals is intuitive and demonstrates how they work in a predictable manner. Using knowledge management processes could make tacit knowledge more widely accessible to everyone within an organisation. In essence, knowledge management processes transform tacit knowledge from something hidden and inaccessible into a valuable organisational asset that can drive performance, innovation, and continuous improvement. The benefits of a strategy or initiative aimed at improving organisational agility and competitiveness in the LNG industry are crucial for project success.

There are many benchmarking organisations but very few that specialise in the benchmarking of mega complex LNG projects throughout their actual phases of the project life-cycles. Given these challenges, organisations specialising in benchmarking mega LNG projects play a crucial role in the industry by providing insights, benchmarks, and best practices that help stakeholders make informed decisions, optimise project performance, and enhance competitiveness in the global LNG market. This is more noticeable in the onshore (downstream) phases of mega complex LNG projects. As such, this is an area that is often misunderstood, and, unfortunately, the accuracy of the benchmarking data may contain errors during these phases of the onshore portion of these large projects. Consequently, there tends to be a cascade effect on the project's overall life-cycle with the lack of accurate phase-based benchmarking in the Assess (Initiate) phase. Therefore, accurate phase-based benchmarking in the Assess phase is essential for making informed decisions, managing risks effectively, optimising costs and schedules, ensuring quality and performance, and driving continuous improvement throughout the project life-cycle. It helps in setting realistic goals, mitigating potential issues early, and ultimately, achieving successful project outcomes. Kerzner (2014; p 48) states, "Even though companies collect best practices, not all best practices are shared outside of the company even during benchmarking studies where all parties are expected to share information." Accordingly, there is a need to use independent companies, or resolute teams, to conduct the benchmarking processes, especially for developing these mega complex LNG projects. This ensures that the normalisation process of the data happens to enable measurements against other projects as like for like, without any preconceived notion of the outcome. Using tacit information, however, does not guarantee an unbiased and true measurement against similar projects. In contrast, explicit knowledge that is systematically documented, validated, and standardised provides a more reliable basis for unbiased measurement and comparison against similar projects. It allows for clearer metrics, benchmarking against industry standards, and more objective assessments of performance, costs, timelines, and outcomes.

Therefore, while tacit information can complement explicit knowledge in decision-making and problem solving, organisations aiming for unbiased and true measurements against similar projects should prioritise the systematic capture, validation, and use of explicit knowledge alongside tacit insights.

Benchmarking on mega/giga complex LNG projects is a means of measuring a project against other successful or completed projects that are already best-in-class and ensuring the complexity factors are visible, such as location, remoteness, etc. Invernizzi et al. (2018) suggest that benchmarking is a valuable tool to improve the performance of projects, but a new framework needs to be considered to deal with the complexities of mega projects.

To achieve this, organisations need to revise and modify their ethos regarding the sharing of knowledge and data into a more open approach with other organisations in the industry. Such a shift in the collective mindset is hard to establish due to the sensitive or protective nature that organisations place upon sharing this type of data. Kaufman (2000) points out that benchmarking is a useful tool; however, he recognises that there is a potential issue with sharing benchmarking data that, if not openly shared and the criteria understood, it can be incompletely or incorrectly used. This lack of openness mostly stems from the majority of organisations unwilling to share with their competitors and details of the successes and failures within the development of their project(s). This would be admitting to others and to JVPs that they got some things wrong. Accordingly, sharing data about projects takes time and requires a great deal of trust and openness among organisations. Currently, as projects end and people move between organisations/projects, they take with them tacit information which they potentially use to develop the next project. Unfortunately, this knowledge (or perceived knowledge) does not always readily transfer to a new/different project and can have detrimental consequences for the project outcome if used unwisely. Cortada and Woods (2000) suggest that there is a danger when managers sometimes try to turn inherently tacit information into explicit knowledge and by doing so can lead to serious problems. A mutually beneficial approach to this issue in the sharing process would be to establish an Australian oil and gas (LNG) benchmarking society, where benchmarking data is freely available and shared openly between organisations. Patty and Denton (2010) suggest that the inherent risk not seen during benchmarking within the same industries is the trust of sharing data between organisations. Drawing from those with experience in benchmarking LNG projects through their life-cycle, this hesitancy to share can be overcome and trust established by oil and gas companies meeting to organise a benchmarking forum where they discuss benchmarking of projects and the benefits of sharing data.

The literature on benchmarking LNG projects, in the form of journals, mostly reviews the risk management of these specific projects. An example is the case study by Stoklosa (1999), which reviews specifically the offshore environmental risks associated with a scenario for the Gorgon project. Other journals, like Jensen (2003), review the LNG revolution, where only two Australian LNG projects are included. When Jensen (2003) benchmarks train sizes, many countries are mentioned, but Australia has not been included in the benchmarking graphs. This journal, however, has excellent data at a very high level and does point out the increasing development of LNG projects around the world, including Northwest Australia. Briggs (2010) points out that 90% of Australia's LNG comes from Northwest Australia. Tusiani and Shearer (2007) observe that the LNG processes in Western Australia (WA) and some other locations have a generic overview of train sizes, start dates, and partner companies. This is an excellent examination of the LNG processes, contracts, etc. However, this could be argued as being far too prominent a level to enable companies to benchmark the life-cycles of their projects. Conversely, McCabe (1998) identifies a need to breakdown benchmarking into the three principal areas, i.e., internal, competitive, and functional. From these bases, organisations can establish the process of benchmarking which is most suitable for their situation.

The main issue in developing projects is that some organisations have the erroneous belief that all projects are alike, which is a complete misconception, particularly in the oil and gas industry and within the rapidly increasing mega complex LNG projects. Shenhar and Dvir (2007) have described this type of constricted thinking as a "project is a project is a project syndrome". With the diversity and complexity of these mega complex LNG projects, they all have their own idiosyncrasies, which determine that no two projects are the same. These consist of different environmental issues, different complexities due to locations, sizes of the LNG onshore plants (i.e., one, two, or three LNG trains – processing units), and the infrastructure to support the development. Shenhar and Dvir (2007) found that, during further research into the reasons for the project failures, projects differ in many ways, and one size does not fit all.

When benchmarking mega complex LNG projects, it is necessary to take an in-depth look at how these projects are evaluated throughout each of the phases (phase-based benchmarking) within their project life-cycles. Throughout each phase, benchmarking provides insights into project performance relative to established standards, identifies areas for improvement, and validates project strategies against successful precedents in the LNG industry. This approach facilitates stakeholders to make informed decisions to enhance project outcomes and mitigate potential risks associated with mega LNG projects.

In this regard, it is also essential to ascertain whether the current conventional project management processes and techniques are suitable enough to manage such large and complex LNG projects. While conventional project management processes provide a solid foundation, managing large and complex LNG projects often requires tailored approaches that account for the unique challenges and dynamics of the industry. Adapting methodologies like agile, incorporating specialised tools for risk management and scheduling, and fostering a culture of innovation and collaboration are essential for successfully managing mega complex LNG projects. According to McKenna et al. (2006) within the oil and gas industry, there is a fundamental change required from the traditional ways project management is used and suggests current ways of doing traditional business is anachronistic.

To achieve success in the future development of LNG projects, there needs to be a major change in basic assumptions in the way organisations and management move forward in managing these projects. In essence, achieving success in future LNG projects requires a holistic approach that integrates advanced project management methodologies, technology adoption, sustainability considerations, and effective leadership. By challenging conventional assumptions and embracing innovation, organisations can navigate the complexities of LNG projects more effectively and sustainably. Shenhar and Dvir (2007) suggest that there is a need to move away from the conventional management processes/techniques used to develop these multifaceted projects that currently follow a standard set of activities outlined in project management books. Accordingly, a new process would need to be established to ensure the correct approach is undertaken for the management of subsequent multidimensional mega complex LNG projects.

The development of mega complex LNG projects in Australia and internationally must ensure the organisational business goals and those of the project managers and project teams are fully aligned. Söderlund et al. (2017; p 7) state, "Megaprojects are built on misinformation about costs, schedules, benefits, and risks. The result is cost overruns, delays, and benefit shortfalls that undermine project viability during project delivery and operations." Consequently, when developing the project life-cycle there is a need to create a flexible approach by using improved problem-solving techniques and open communications. These need to be included within the teams' philosophies (KPIs and BSCs) during each phase of the project life-cycle going forward. Kaufman (2000) suggests that if you can name it, you can measure it, pointing out that we can (should) measure everything. This means the project team ought to ascertain the need for a re-evaluation of both schedule and costs as the project enters each phase, including activity-based and the phase-based benchmarking processes. This would provide visibility with the dynamic scope of works that these complex projects encounter in each phase. Akintoye (2010) points out that project complexity is a factor that can hinder performance on site in the construction phase including the unpredictability of the work on site, with the scale and scope of construction complexity of the design.

Unfortunately, even despite all the rigorous checks and balances at each phase or and risk gates, at the end of every phase, of these extraordinarily complex LNG projects, a trend identified by the author seems to occur which reveals the undefined Scope of Works issues emerge after the FID. Even though, significant investments are made based on agreed project scopes and timelines, and any undefined or evolving scopes can lead to cost overruns, delays, or project scope creeps. Addressing this issue requires careful planning, clear communication, and detailed documentation during the pre-FID phase (Develop). Early and comprehensive risk assessments and robust contracting strategies (e.g., defining deliverables clearly) are essential to minimise scope ambiguity later in the project life-cycle. Additionally, adaptive project management strategies provide the flexibility needed to handle evolving and undefined work throughout the project life-cycle. By embedding flexibility into the project execute phase, teams can better navigate scope changes, minimise cost overruns, reduce delays, and maintain alignment with project objectives. This juncture could be referred to as the 'point of no return' or Go/No-Go decision-point' is a critical juncture in project management, particularly in large-scale, high-investment projects like LNG developments. Any changes or unforeseen scope creep beyond this stage will inevitably lead to substantial schedule delays and cost overruns. Criaud and Boulinguez (2013; p 10) state, ". . . the potential functional confounds in go/no-go paradigms, do not only concern the pre-frontal cortex. They also involve posterior regions in the parietal and temporal cortices."

To achieve better outcomes for their projects, management needs to embrace a new ethos of change to move away from the conventional management approach to developing the life-cycle of mega complex projects. Söderlund et al. (2017; p 8) state, "By emphasizing the importance of what goes on in projects, we are moving beyond the conventional approaches to what goes on before megaprojects are launched or after they are completed." Part of this change is to remove the constant pressure in the Develop phase from corporate and JVP management to reach FID as soon as possible, consequently reducing or holding the FID date even though more scope may be identified after FID. By doing so, this constraint does not allow enough time for the project team to understand and make visible any additional scope of works prior to FID.

With the intention of developing a different management structure for their mega/giga complex LNG projects to be more successful, there should be a prerequisite for organisations to encompass a mindset change regarding how they should manage these projects. Elmuti and Kathawala (1997) suggest that the benchmarking process for projects should provide the basic framework for action, with flexibility for modification to meet individual project needs. This approach would be for the benchmarking of the durations of the critical phases (phase-based benchmarking) of the project life-cycle, such as the Develop and Execute phases. According to Akintoye (2010), the supervision requirements on-site are critical to the efficiency of the construction, which may require a re-organisation of the company supervisors and existing labour forces, according to the project team experience, to match the company expectations of success. As a benchmarking process within the project life-cycle, when the project reaches the Develop phase would be an opportune time to adopt a new structure of management to proceed into the Execute phase. Akintoye (2010) suggests that the decision for a re-organisation needs to be considered within the overall corporate business strategy of the organisation. To achieve this, the review should commence with the management hierarchy structure to ascertain how this would be incorporated from the commencement of the Develop phase going forward. This ensures that the hierarchy is structured to remain in place from the Execute phase through to the RFSU. The management structure should also have provision for a corporate compliance director to be included in the project management team. According to Andersen et al. (2009), the influences from the base organisation are stronger when people work simultaneously for the project and the overall organisation (matrix organisation) than when they concentrate all their efforts on the project alone. With this proposed modification in the management structure, it then provides a timeframe to benchmark the competency requirements for senior management to accomplish before they are able to undertake the management of these multifaceted LNG projects. Turner and Muller (2005) suggest that the dilemma that organisations often face is between keeping individuals working on the current project or moving them to career opportunities that arise in the same period. At present, most organisations change managers at the end of each phase of these projects, which tend to diminish the overall ownership and accountability throughout the project life-cycle. Merrow (2011; p 123) states, "... governance is about the accountability of management for the health and performance of the corporation." Managers and some team members may leave after completing their life-cycle phase without recourse to any issues or problems they may have inadvertently or unwittingly caused for the next phase of the life-cycle. Keegan and Turner (2000) point out that this tends to be recognised as the no home syndrome because of the project durations in the phases people tend to move quickly between projects. It could be argued that any such issues or problems with people moving from the project should be part of the phase-gate risk process before entering the next phase. By incorporating turnover related risks into the phase-gate risk management process, organisations can proactively mitigate potential disruptions caused by personnel changes. This approach not only enhances project resilience but also contributes to overall project success by ensuring continuity of expertise and minimising the impact of turnover on project outcomes. Unfortunately, such soft issues are not usually considered as a major project risk by some organisations.

It is at this juncture where organisations should allocate sufficient time to undertake strategic benchmarking of their projects, including their commercial aspects, which is crucial at key junctures such as phase gates. To effectively undertake strategic benchmarking of mega/giga LNG projects, organisations should allocate dedicated time and resources, involve relevant stakeholders (including project managers, finance experts, and commercial advisors), and leverage external expertise and industry benchmarks where applicable. This systematic approach not only enhances project management practices but also contributes to long-term project success and sustainability in the competitive LNG market. According to Lankford (2000), strategic benchmarking deals with top management and how they need to look at the long-term results of the company and the projects, such as natural gas (NG) or liquefied natural gas (LNG) projects, to make them successful.

4.5 Project Life-Cycle Time and Complexity Framework

Mega LNG projects from conception (Assess phase) to completion (Execute – RFSU) can be a ~6 to 10-year investments for stakeholders. As identified in Chapter 4 Section 4.3 'Research Approach Justification', there are five main phases to the project life-cycle of LNG projects. Each of these phases has benchmarked durations, as part of the overall framework which need to be reviewed during the phases, as the project methodology definition is better refined. This occurs prior to the operate/evaluate phase, which is a ~3-month period, where the project is combined with operations for the evaluation and final hand-over for first contract gas and close-out of the CAPEX budget.

Consequently, for mega/giga LNG projects, it is envisioned that the life-cycle can extend over long periods from the Assess phase (conception) to the Execute phase (completion – RFSU), depending on how many process trains are being constructed and the complexity associated with them. Haidar and Ellis (2010) suggest that the number of different elements within a project and their interrelatedness can determine complexity. Katebah et al. (2020; p 2) state, "The LNG processing train is divided into two subsections: hot and cold. As the name indicates, the hot section operates at above ambient temperatures and it encompasses the NG pre-separation, sweetening, SRU, and dehydration units. On the other hand, the cold section is at sub-ambient conditions and includes the natural gas liquids (NGL) recovery and fractionation, liquefaction using propane/mixed refrigerant cycles (C3MR), helium extraction (HeX) and nitrogen removal (NR) units, and the export terminal."

Accordingly, when the economy of scale is to construct an LNG plant with three trains, the stakeholder investment can be ~10 years from Assess to RFSU. As these projects evolve from mega to giga, the resultant project complexity increases. Haidar and Ellis (2010; p 4) state, "Theoretically, a mega project is not different than any other construction project. It includes numerous parallel activities, limited resources, tight schedules, and multiple decision-making parties." However, they have not considered the multifaceted aspects of developing giga LNG projects, which are far more complex in design, JVP intervention, and managerial requirements as they proceed throughout their project life-cycles. Aritua et al. (2009; p 72) state, ". . . the need for new approaches, processes, and techniques suitable for multi-project management is obvious. This, however, requires a change in mindset . . ."

4.5.1 Phased-Based Benchmarking

Phase-Based Benchmarking (PBB) is an emerging approach that focuses on evaluating and comparing the performance of projects during different phases of their life-cycle. Unlike traditional project management systems, which often provide overall performance metrics, PBB introduces a set of performance metrics that are specifically tailored to assess project performance at various stages or phases. These metrics are designed to provide a more granular phase-specific view of assessing project performance, allowing project managers to identify strengths and weaknesses at different stages of the project life-cycle. While traditional project management systems may not explicitly track these phase-based metrics, they can be integrated into project management practices through custom reporting or specialised project management software that supports PBB. Unlike traditional metrics such as time, cost, and scope, PBB focuses on evaluating performance at different stages or phases of a project life-cycle. These PBB metrics develop a unit of analysis at the project phase-level. As described in Chapter 6, 'LNG projects and Performance Measurement', the current units of analysis for benchmarking mega complex LNG projects in the phases are inadequate and do not provide consistent or reliable performance measures for the size and complexity of each of the project phases, as described in Figure 4.1 – 'Project Complexity Framework'. The use of phase-based performance metrics allows the owners and project managers to evaluate project performance at the phase-level to allow proactive strategies to be addressed (Yun et al. 2016). Such performance metrics enables project managers to assess performance outcomes in all phases of a project life-cycle and enable the evaluation of each phase against similar projects for improving benchmarking at this level for future project developments. Fong et al. (1998; p 416) state, "Once benchmarking parameters of change are identified, the benchmarking team should integrate the findings into the organization, including sharing the idea of change with those who would support and provide input into the process and those who will be affected by the changes."

4.6 Upstream

Upstream benchmarking for structures, such as jackets and topsides, are shared amongst oil and gas organisations through benchmarking houses such as the Performance Forum (Edinburgh) and Independent Project Analysis (IPA), compared to the downstream (onshore) portion of an LNG project. Mostly this data comes in the form of scattergrams for cost and durations, as illustrated in the tables below.

These scattergrams are broken down into phases of topsides and jackets (steel structures or Jack-ups, etc.) from design and fabrication for both cost and schedule and overall sanction to RFSU. The graphs shown in Tables 4.1 and 4.2 are reproduced from the data obtained from the Performance Forum's examples of Topsides total design comparison in the duration of months.

Table 4.1 Performance Forum Topsides by Weight

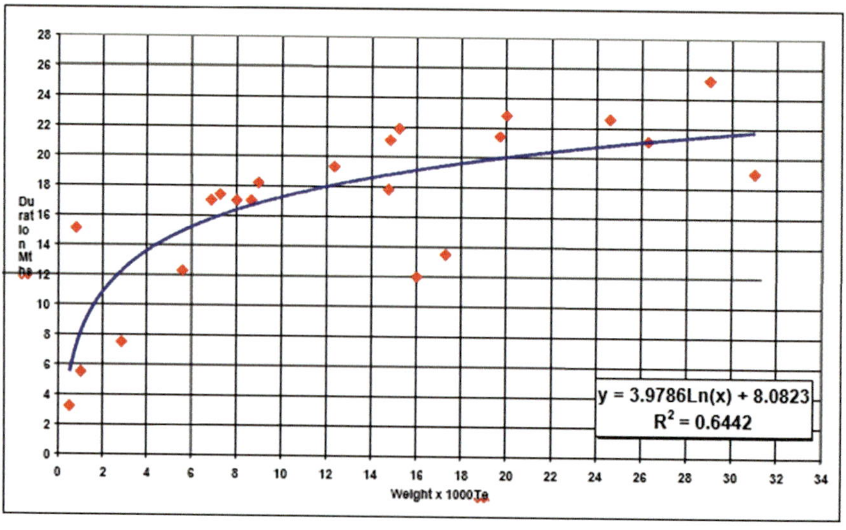

(www.performance-forum.com)

Table 4.2 Performance Forum Topsides Fabrication Duration

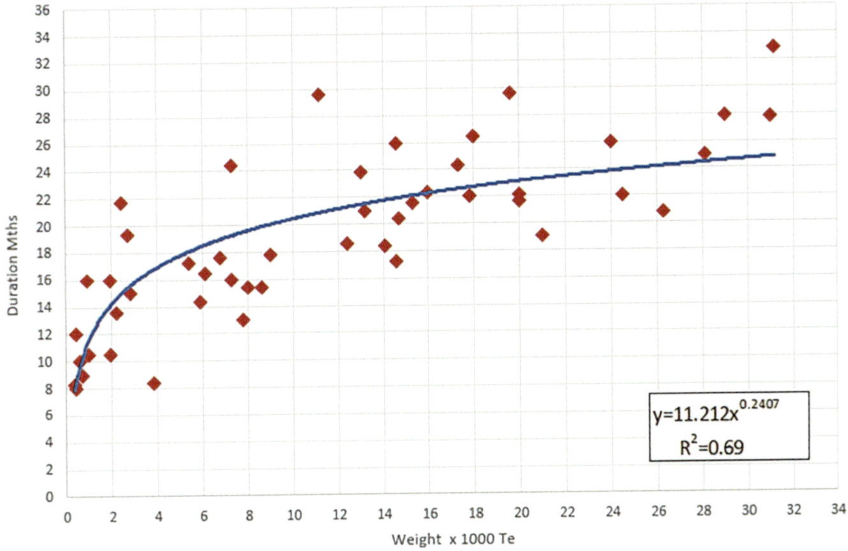

(www.performance-forum.com)

For clarity R^2 as shown on Tables 4.1 and 4.2 represents:

- "R^2 (R-Squared) is a measure of how well a linear regression model fits a dataset."

In technical terms, it is the proportion of the variance in the response variable that can be explained by the predictor variable.

The value of R-squared can range from 0 to 1:

- A value of 0 indicates that the response variable cannot be explained by the predictor value at all; and
- A value of 1 indicates that the response variable can be perfectly explained without error by the predictor variable.

In practice, you will likely never see a value of 0 or 1 for R-squared. Instead, you'll likely encounter some value between 0 and 1.

Due to the confidential nature of this data, these independent companies do not divulge on the graphs the project or company names within the illustrated data sets they have received from various organisations.

From these benchmarking graphs, organisations can select the weight of their topsides and predict the design duration, which should be ranged around a +/- 3-month window to allow for different contracting strategies the projects may use.

The dry weight of the topsides is available as benchmarking data for the topside fabrication durations. To ensure that there is flexibility within the benchmarking, the use of ranges to be used around the weight are projected in the region of +/- 6 weeks.

Table 4.3 Shell FLNG Compared to Major Structures

Comparison Figure	
Description	*Metres*
Taipei 101 Tower	508
Prelude (Shell FLNG)	**488**
Petronas Towers	452
Eiffel Tower	325
N Seoul Tower	236

Source: Author's Research

4.6.1 FLNG

This book does not include details of Floating LNG (FLNG) but is worthy of a brief overview of the largest FLNG project in the world, i.e., Shell's Prelude. This vessel is the largest vessel to be in production for LNG and is situated in the Northwest of Australia.

According to Sakmar (2013), Prelude is the largest offshore FLNG facility ever built, measuring 488 m long and 74 m wide. Reference to Table 4.3.

As it is the only FLNG processing vessel of this size and kind, it is not possible to benchmark this type of process facility until qualitative data becomes available as more similar floating LNG process plants are developed.

4.7 LNG Projects and Benchmarking

Based on the performances of mega complex LNG projects in Australia and worldwide, it is evident that the current project management processes need to be updated, particularly the way companies benchmark each phase (phase-based benchmarking) of a project life-cycle. While traditional phase-based benchmarking has its merits, updating project management processes for mega LNG projects in Australia (and globally) should involve a more tailored and dynamic approach. These large, complex projects often face cost overruns, delays, and operational inefficiencies, which are sometimes linked to outdated or rigid benchmarking processes.

By moving away from rigid phase-based benchmarking and adopting a more integrated, flexible, and dynamic approach, LNG projects in Australia (and globally) could see improved outcomes in terms of cost control, schedule management, and overall project success. This includes developing customised benchmarks, integrating advanced technologies, fostering collaboration, and maintaining a strong focus on risk management throughout the project life-cycle. Customisation ensures that benchmarking is aligned with the unique characteristics of each project, stakeholder priorities, and changing market conditions, instead of relying on generic or one-size-fits-all metrics. Tusiani and Shearer (2007; p 132) state, "The critical path for construction of an LNG plant is usually driven by the delivery of long-lead items, such as heat exchangers, compressors, large turbines, and LNG storage tanks." Whereas Mokhatab et al. (2014) suggest that the LNG supply chain extends to upstream production, LNG production plants (usually onshore), shipping, storage, and regasification to sales gas. The change in benchmarking processes for developing large LNG projects underscores the importance of reassessing the duration of the project life-cycle. Traditionally, the development of such projects involves multiple stages that can span several years, typically including exploration, engineering design, regulatory approvals, construction, commissioning, and finally, operations. Accordingly, more emphasis needs be on the Develop and Execute phases where good quality decision-making is crucial due to the scale and complexity of reserves or reservoirs involved. Subsequently, emphasising quality decision-making in the Develop and Execute phases of mega LNG projects is pivotal for managing the inherent complexity and scale of these endeavours. By leveraging data-driven insights, fostering collaboration, managing risks effectively, and ensuring robust governance, project stakeholders can enhance their ability to deliver successful outcomes aligned with the size and scope of the reserves or reservoirs being developed. The Develop phase of the project is a critical period where pressure comes from some executives and JVPs to complete this phase as quickly as possible to enable the project to attain FID. The author's observation on this phase is that these demands are normal and, as such, prominent during the project development. Acknowledging that these demands are normal and prominent underscores the industry's recognition of the inherent challenges in developing large LNG projects. Addressing these demands effectively requires a comprehensive understanding of the industry landscape, proactive planning, robust project management practices, and continuous adaptation to changing market dynamics and regulatory requirements. According to Merrow (2012), schedule pressure dooms more mega projects than any other factor due to organisations wanting to move a project along quickly.

During the development of a mega complex LNG project, there is a point, which is at the commencement of the Develop phases (depending on the size/complexity of the project), where the project is divided into two distinct segments of Upstream (offshore) and Downstream (onshore). These two segments are managed as standalone projects; however, they do have multiple interfaces that are aligned during the Develop and Execute phases to ensure they come together to attain the RFSU. These enormous LNG projects are becoming so complicated that there tends to be a problem with benchmarking them accurately. To address these challenges, stakeholders in the LNG industry often use a range of metrics and benchmarks tailored to specific project parameters and contexts. These could include cost per ton of LNG produced, construction timelines, operational efficiency metrics, and financial returns adjusted for risk factors. Despite the complexities, benchmarking remains crucial for assessing project performance, identifying areas for improvement, and making informed investment decisions in the LNG sector. The upstream (offshore) side, which includes the sub-sea systems (pipeline to onshore) interface, also known as the beach valve, can have particularly good benchmarking data available for use. However, there seems to be a consistent dilemma occurring on how to accurately benchmark the downstream (onshore) portion of the processing plant, causeways, MOF, jetty, infrastructure, and environmental requirements, as sharing of benchmarking data is not a practice that happens openly between organisations.

With the economy of scale of capital expenditure on projects, it is imperative to develop the life-cycle methodology for benchmarking these complex LNG mega/giga projects. By implementing a robust life-cycle methodology for benchmarking, stakeholders in LNG mega or giga projects can better manage costs, timelines, risks, and performance metrics. This approach not only enhances project transparency and accountability but also supports informed decision-making throughout the project life-cycle. Keehley and MacBride (1997; p 78) state, "Benchmarking has evolved from a nebulous concept into a concrete method used to identify and import best practices."

One factor that can disrupt these massive LNG projects is the lack of recognition of the high impact/high-risk factors believed to be so incredibly unlikely that they are discarded in the early phases (Assess or Select) or not brought to the forefront at all. These types of phenomena are known as the outliers or black swans, which are the unexpected events that do happen on projects. They are characterised by their rarity, significant impact, and the challenge they pose to conventional expectations or planning. These phenomena can range from sudden changes in requirements, unexpected technical issues, or even external factors like natural disasters or economic shifts. Managing outliers or black swans often requires contingency planning, flexibility, and the ability to adapt quickly to new circumstances. Raydugin (2012) suggests that a major problem with projects is the lack of risk estimation for the known unknowns and the unknown unknowns (black swans or outliers).

However, during brainstorming exercises at the development of projects, the outliers are dismissed as being too extreme to even be considered or not even mentioned at all at this stage. The main reason for this disregard is usually due to the fact that the majority of people who participate in brainstorming sessions or are part of the team are reluctant to mention these due to the possibility of being mocked or ridiculed by their peers. This fear of social judgement can stifle creativity and prevent innovative ideas from being explored.

The possibility and significance of outliers or black swans needs to be discussed in-depth with JVP representatives when undertaking these complex LNG projects. According to Rolstadas et al. (2011), black swans are outliers and are the risks that are usually ignored until they occur with devastating impacts, usually in the implementation/Execute phase. It is for these reasons that a new ideology ought to be adopted for the management of these projects when undertaking these massive developments. Remington and Pollack (2012; p 36) state, "A number of traps need to be watched for in structurally complex projects. There is often inadequate attention to detail in the early phases of the project, resulting in a poor estimation of time and cost." Consequently, outside-the-box approaches need to be considered in the initial stages or phases of the project life-cycle. Therefore, incorporating outside-the-box approaches in the initial stages requires a supportive environment where team members feel encouraged to voice unconventional ideas without fear of criticism. Potential black swan events need to be highlighted and not ignored in these earlier stages of setting up mega complex LNG projects. Taleb (2004) implies that black swans are a random event which can contain three main properties: substantial impact, surprise effects, and incomputable probabilities.

Black swans are more prevalent in mega complex LNG projects as these projects are so complicated. Consequently, in the earlier development of the project life-cycle, there is usually insufficient information to define the Scope of Works for benchmarking. According to Raydugin (2012), unknown unknowns are what the project team did not identify as uncertainty objects and, as such, are overlooked or hidden as unknown unknowns. These high impact risks (black swans) need to be visible or at least entered into the project risk register as considerable risk.

Consequently, they should not be discarded but the cause and effects of these occurring monitored throughout each phase as the project matures. These risks can also affect how the project is estimated at the beginning and throughout the first three phases. Andersen et al. (2016) suggest that projects experience escalating costs throughout the various stages, which can be owing to underestimating the budget at the front end to keep the budget as low as possible to move the project forward.

With mega complex LNG projects, it is crucial that the emphasis is on the duration of the Develop phase of the project life-cycle to allow for better scope definition prior to FID. Subsequently, prioritising the duration and thoroughness of the Develop phase in mega LNG projects is crucial for achieving better scope definition prior to the Final Investment Decision. It lays the groundwork for successful project execution by minimising risks, optimising costs, adhering to schedules, and fostering stakeholder alignment. For example, instead of allowing a duration of ~18 months in the Develop phase, the duration should be benchmarked at ~24 months to allow more in-depth studies to provide a better definition of the Scope of Works. This would provide a more robust scope prior to the final risk gate that moves the project to FID and then into the Execute phase. This upfront additional expense goes a long way to mitigating the excessive cost and schedule overruns that only become apparent in the Execute phase. This is when scope creep begins due to the inadequate definition of the scope of works in the Develop phase and lack of benchmarking due diligence. Merrow (2012) implies that skimping on the front-end definition of a project is foolish, as senior managers can find themselves in a dilemma because they do not understand the cost and scope of the project at the appropriate time. Andersen et al. (2016) suggest that, in the concept phase, people who want the project to be approved keep the first cost estimate optimistically low; this is known as anchoring. As suggested above, with the extension of the Develop phase to ~24 months, it provides the additional time required to ensure the Scope of Works is fully visible to the team.

Consequently, the more time spent in getting this phase correct, the less potential there is for cost and schedule overruns in the Execute phase. According to Merrow (2012), unattainable speed in mega projects destroys these mega projects and is responsible for sheer vaporisation of hundreds of billions of dollars and the ruination of project outcomes.

For project managers and organisations to accept this period increase, it would take a quantum frame of mind-change in organisations' management processes. This would also require the management team of the execute phase to collaborate closely with the Develop team during the Develop phase.

The author's observations suggest that one of the main paradigm shifts that should occur in giga LNG projects is to move away from the competitive tendering process and work on long-term partnerships as these project life-cycles extend for ~6–10 years or even 15 years, depending on the complexity of the project being developed. By doing so, it promotes collaboration, risk-sharing, cost efficiency, innovation, flexibility, and stakeholder engagement, ultimately contributing to the successful execution and sustainable operation of these complex projects over extended life-cycles. The view from the Egan (1998) report was that competitive tendering should be replaced with long-term relationships.

Consequently, the author's observations suggest that there is a reluctance for some organisations to acknowledge the issues they face when managing complex LNG projects. They could be unwilling to accept that the fundamental processes being utilised on conventional projects, such as Lean Sigma and Lean Construction, no longer provide a sustainable outcome when dealing with these mega complex LNG projects. This is due to their complexity and long durations compared with smaller projects. According to Raydugin (2012), the Six Sigma approach was originally from operations, not from projects. This statistical approach requires repetitiveness to promote frequency-based evaluations. This premise that such lean approaches work on mega complex LNG projects continues to prove unreliable as these projects constantly overrun on cost and schedule when trying to use the lean approach with such multifaceted project scopes. For organisations to move away from these processes would be a significant challenge in applying lean approaches to mega LNG projects, which often have multifaceted scopes and face substantial complexities. Therefore, stakeholders in mega LNG projects may need to adopt a hybrid approach that incorporates elements of lean thinking alongside robust project management practices, risk management strategies, and stakeholder engagement frameworks. By doing so, this approach can help balance the need for efficiency with the realities of complexity, uncertainty, and scale inherent in these ambitious ventures.

4.8 Benchmarking Mega Complex LNG Projects

Benchmarking as a performance measurement of mega complex LNG projects has been an important strategy for oil and gas organisations around the globe. As such, benchmarking plays a pivotal role in the oil and gas sector, particularly for mega and complex projects like LNG developments. It serves as a strategic tool for performance measurement, identifying best practices, improving efficiency, managing risks, fostering continuous improvement, and enhancing stakeholder confidence. By leveraging benchmarking insights, organisations can enhance their competitiveness and achieve successful outcomes in challenging project environments. Horváthová et al. (2021; p 3) state, "Benchmarking studies can be divided according to the type of benchmarking into: process, functional and performance benchmarking studies."

To significantly contribute to the literature on benchmarking, it is necessary to understand the concept of benchmarking itself as a performance measurement and how it relates to specific oil and gas projects or complex LNG projects. Consequently, in this section, the definition, benefits, types, structure and the project life-cycle are presented. The way organisations use the performance measurement system, including the use of Balanced Scorecards (BSCs), Earned Value Management Systems (EVMS), and key performance indicators (KPIs), is also discussed.

4.8.1 Shortcomings on Managing and Benchmarking Complex LNG Projects

There are significant shortcomings within the current process when trying to benchmark these complex LNG projects, including management direction and support to provide funds to develop benchmarking frameworks. According to the author's observation, addressing shortcomings requires proactive leadership and strategic planning from senior management within oil and gas organisations. They must prioritise investment in benchmarking initiatives, provide adequate funding and resources, foster collaboration with industry peers for data sharing, and empower teams with the necessary expertise to develop and implement robust benchmarking frameworks. By overcoming these challenges, organisations can derive valuable insights, drive continuous improvement, and enhance the overall performance of complex LNG projects. Limbrick (2009; p 14) states, "The lack of senior management support or buy-in erodes the value of project benchmarking." One of the key issues is the resistance of oil and gas organisations to share their project data (benchmarking), such as phase durations/costs, among one another when developing these multifaceted projects. Horváthová et al. (2021; p 3) state, "Benchmarking is also an important tool for identifying key performance indicators. Well-known key performance indicators are related to financial analysis." Previously, Andersen and Pettersen (1996; p 112) state, "Benchmarking is very much about creating an environment or network of companies where exchanging information for comparison is accepted." The author's observations suggest that most, if not all, organisations continue to believe that they can bring these massive projects in on budget and schedule using conventional project management techniques and processes or tacit information and ignore benchmarks of previous smaller projects. Haidar and Ellis (2010; p 1) state, "Limited research has been done in the construction industry to investigate megaprojects and to improve their performance."

There are times when organisations believe their project is so unique or specialised that it cannot be effectively compared to any other project for benchmarking purposes. This could be due to various factors such as location, scale, technical requirements, or other specific characteristics that set it apart from other projects. When a project is deemed "special" in this way, it often means that conventional benchmarks or industry standards may not fully capture its complexities or challenges. In such cases, it becomes crucial for project managers to develop customised metrics and evaluation criteria tailored to the project's unique circumstances. However, it is also important to ensure that this assertion doesn't become a shield against accountability or improvement. While recognising uniqueness is valid, it shouldn't preclude efforts to learn from past experiences or industry best practices that might still offer valuable insights or lessons applicable to the project at hand.

Managing mega/giga LNG projects presents unique challenges that often require a departure from conventional project management processes. According to the author's observations, understanding why these traditional approaches fall short and exploring alternative strategies for success, such as:

1. **Complexity and Scale**:
   - Mega/giga LNG projects involve massive investments, intricate technology, and long-lasting impacts on the economy, environment, and society;
   - Traditional project management methodologies may struggle to address the sheer complexity and scale of these ventures.
2. **Remote Locations**:
   - Many liquefaction plants are situated in remote areas, making logistics and resource allocation challenging;
   - Conventional processes may not adequately account for the specific requirements of these locations.
3. **Diverse Workforce and Cultures**:
   - LNG mega/giga projects involve global collaboration, with diverse teams, vendors, and financial institutions;
   - Traditional approaches may not effectively manage cultural differences and communication challenges;
   - While traditional project management approaches may have their strengths, they often fall short in effectively managing cultural differences and communication challenges in global mega projects like LNG developments;
   - Embracing modern practices that prioritise cultural intelligence, inclusivity, and adaptability can foster more effective collaboration, enhance project outcomes, and mitigate risks associated with cultural diversity.
4. **Unclear Scope**:
   - Acquiring lump-sum contracts without a well-defined scope can lead to problems;
   - Project managers must ensure clarity, proper communication, and agreement on project scope;
   - It is essential for project managers to successfully execute any project, especially complex ones like mega complex LNG developments.
5. **Adaptive and Agile Approach**:
   - Given the continuously changing needs of LNG mega/giga projects, project management teams must be adaptive and agile;
   - Rigid adherence to traditional project management methods may hinder project success;

- Project managers should leverage adaptable approaches, such as agile methodologies, integrated project management frameworks, and collaborative decision-making processes, to navigate uncertainties, foster innovation, and achieve sustainable project outcomes in the evolving LNG industry landscape.

However, conventional project management processes, which may work adequately for smaller LNG projects, often face significant inadequacies when applied to mega/giga LNG projects, due to their multifaceted and complex structures. Project managers involved in mega/giga LNG projects must embrace adaptability, clear communication, and a holistic understanding of the project's unique context, which are critical factors for ensuring successful execution. This enables them to navigate complexities, manage risks, foster collaboration, and drive continuous improvement, ultimately delivering value and achieving sustainable project outcomes. McIntosh (2018; p 2) states, "The entire Project Execution – from Initiation to Execution and Close-Out must follow a rigid process that ensures the requisite amount of Front-End Loading (FEL) is applied before – and is, in fact, a condition for – Project Sanction."

Collectively, using conventional PM processes, there appears to be a consistent miscommunication regarding the time it takes to build the infrastructure and to develop a robust supply chain. According to the author's observations, it is proposed that the infrastructure needs to be part of the pre-investment, prior to FID. By integrating infrastructure planning into the pre-investment phase before FID is critical for mitigating risks, optimising costs, managing timelines, engaging stakeholders effectively, ensuring operational readiness, and enhancing overall project feasibility in LNG developments. By addressing infrastructure needs early, organisations can lay a solid foundation for successful project execution and long-term operational success.

The fabrication of these multifaceted mega/giga LNG projects is contracted to numerous local and international entities. This multifaceted approach adds significant complexity to equipment completion and delivery to site. This is due to the fact that no single fabrication shop can deal with the size and timeframes required for delivery of modules, PAUs, PARS, and multiple piping arrangements. Coordination among various contractors and suppliers, both domestic and international, becomes crucial to ensure timely fabrication, transportation, and installation of equipment required for the LNG facilities.

As a result, the fabrication process spans across multiple countries intensifying the strain on the supply chain responsible for delivering various sections of the project within scheduled timeframes. This global distribution of fabrication tasks necessitates meticulous planning and efficient management to ensure that all parts arrive on schedule and meet the stringent quality and safety standards required for such large-scale LNG projects.

This subsequently adds additional strain on the supply chain for the delivery of the many sections of the project to meet schedule timeframes. Whilst managing a global supply chain may present challenges, strategic planning, effective communication, and the adaptation of technology can help ensure timely delivery and project success. Developing a comprehensive supply chain strategy that includes contingency plans, risk management strategies, and clear timelines is essential. This involves mapping out the entire supply chain from sourcing raw materials to final delivery and installation. By employing robust strategies, project managers can navigate the complexities of global supply chains more effectively, ensuring timely delivery of components and ultimately achieving project success for multifaceted LNG projects.

4.8.2 Future of Mega Projects Life-Cycle Benchmarking

When developing mega/giga LNG projects, organisations need to reflect on the best means of benchmarking these multifaceted LNG projects. They could, as a current practice, use reputable benchmarking houses to do this for them. However, it could be argued whether or not they provide the full scope of works to the benchmarking houses to confirm the project complexity factors, which include updating the scope of works as the project moves forward throughout the phases to achieve FID. According to Patanakul and Shenhar (2012), a new mindset and approach must be built on top of the traditional project management concept. Consequently, there is a need for a revised means of using benchmarking strategies and techniques for mega/giga complex LNG projects moving forward. Accordingly, mega and giga LNG projects require a significant shift in mindset and approach compared to traditional project management. This demands a blend of strategic foresight, technological agility, rigorous risk management, and effective stakeholder engagement. It involves project managers to go beyond traditional methodologies and embraces innovative solutions tailored to the unique challenges of mega/giga LNG developments on a massive scale. These projects are often highly complex, involving numerous stakeholders, intricate technical requirements, significant investments, and long-term commitments. As such, benchmarking strategies and techniques must be adapted to suit their unique characteristics. Revising benchmarking approaches can be crucial for maintaining relevance and accuracy. Making the benchmarking process more relevant, effective, and aligned with the organisation's strategic goals is a crucial process. Some considerations for revising benchmarking methodologies are as follows:

- **Customised Benchmarks**: Traditional benchmarks may not adequately capture the complexities of mega/giga LNG projects. Customised benchmarks tailored to the specific requirements, challenges, and objectives of these projects are essential. This might include benchmarks related to safety performance, environmental impact, project timeline, cost efficiency, and operational excellence.
- **Holistic Metrics**: Mega/giga LNG projects involve multiple dimensions beyond just cost and schedule. Benchmarking efforts should encompass a holistic set of metrics that reflect the project's overall success criteria, such as technical performance, regulatory compliance, stakeholder satisfaction, and long-term sustainability.

- **Industry Collaboration**: Given the scale and complexity of mega/giga LNG projects, collaboration within the industry can provide valuable insights and benchmarking data. Establishing industry consortia or partnerships for sharing best practices, lessons learnt, and benchmarking information can help raise performance standards and drive continuous improvement across projects.
- **Long-Term Perspective**: Mega/giga LNG projects typically span several years from conception to operation. Benchmarking strategies should consider the long-term life-cycle of the project, including pre-FID (Final Investment Decision) stages, construction, commissioning, and ongoing operations. Evaluating performance against benchmarks at each stage (phase) can help identify areas for optimisation and mitigate risks of cost overruns or delays.
- **Risk-Adjusted Benchmarks**: Mega/giga LNG projects often face inherent risks related to geopolitical factors, market dynamics, technological challenges, and environmental considerations. Benchmarking techniques should account for these risks and incorporate risk-adjusted metrics to assess project performance more accurately.
- **Continuous Learning and Adaptation**: Benchmarking should not be viewed as a one-time exercise but as an ongoing process of learning and adaptation. Project teams should actively monitor performance against benchmarks, identify trends and deviations, and implement corrective actions in real-time to stay on track towards project objectives.

By adopting these proposed benchmarking strategies and techniques, stakeholders involved in mega/giga LNG projects can significantly improve their ability to assess performance, identify areas for improvement opportunities, and ultimately enhance better project outcomes.

Adapting this approach change can be quite challenging, especially considering the entrenched nature of the current expertise in developing large-scale LNG projects. This is specifically true for those with extensive project management experience and, accordingly, there tends to be a resistance to change which, in most cases, is hard to implement. By approaching the transition to modern benchmarking methodologies systematically and strategically, stakeholders can gradually overcome challenges and achieve better outcomes in mega/giga LNG project developments. Koc-Menard (2009; p 31) states, "Many companies find it attractive to employ 55-year-old – plus workers because they bring in valuable assets, such as long experience, in-depth organizational and technical knowledge and rich networks of professional contacts." With experienced knowledge leaving the oil and gas industry, it is essential that benchmarking data (including tacit information) for the future is captured and preserved. Changing mindsets and overcoming resistance to change, especially in industries characterised by established practices and experienced professionals, can be a formidable challenge. Some strategies to address the resistance to adopting new approaches in mega/giga LNG projects are as follows:

- **Education and Awareness**: Commence by educating key stakeholders about the need for change and the benefits of adopting new approaches. Highlight the evolving dynamics of the LNG industry, emerging trends, and the potential risks of sticking to outdated practices. Use case studies, industry reports, and real-world examples to illustrate the advantages of embracing innovation and modernisation.

- **Engage Leadership Support**: Secure buy-in from senior management and influential leaders within the organisation. Leadership support is crucial for driving organisational change and overcoming resistance. Convey the strategic importance of adapting to new market realities and positioning the organisation for long-term success.
- **Demonstrate Success Stories**: Showcase success stories and best practices from other projects or organisations that have successfully implemented innovative approaches in mega/giga LNG projects. Highlight tangible results, such as improved efficiency, cost savings, enhanced safety, or accelerated project timelines, to inspire confidence and demonstrate the potential benefits of change.
- **Collaborative Approach**: Involve key stakeholders, including experienced project managers in the change process. Seek their input, feedback, and involvement in designing and implementing new strategies. By engaging them as partners rather than adversaries, you can leverage their expertise while fostering a sense of ownership and commitment to the change initiative.
- **Pilot Projects and Incremental Changes**: Introduce new approaches gradually through pilot projects or phased implementations. Start with small-scale initiatives that allow stakeholders to experience the benefits firsthand and build confidence in new methods. As success is demonstrated, gradually scale up the adoption of new practices across larger projects.
- **Provide Training and Support**: Offer training programs, workshops, and resources to equip project teams with the knowledge and skills needed to implement new approaches effectively. Address any concerns or gaps in understanding through targeted education and ongoing support mechanisms.
- **Create a Culture of Continuous Improvement**: Foster a culture that values innovation, learning, and continuous improvement. Encourage open communication, collaboration, and a willingness to challenge the status quo. Recognise and reward individuals and teams that embrace change and contribute to driving positive outcomes.

By employing these strategies, organisations can begin to overcome resistance to change and facilitate the adoption of new approaches in mega/giga LNG projects, paving the way for greater success and competitiveness in the evolving LNG industry landscape.

Establishing and maintaining the future of benchmarking processes in the oil and gas industry, particularly in organisations focused on mega/giga LNG projects, requires a forward-thinking approach that addresses evolving industry dynamics, technological advancements, and changing stakeholder expectations. Baek and Lee (2009; p 274) state, "The practice of benchmarking is critical to Total Quality Management (TQM) and re-engineering. To measure the effectiveness of TQM and re-engineering, firms must assess their performance against their benchmarks."

While traditional project management concepts of cost and schedule are fundamental, they are insufficient for effectively benchmarking mega/giga LNG projects. Modernising benchmarking processes to incorporate a broader range of metrics, embrace technological advancements, enhance risk management strategies, and ensure regulatory compliance is essential to achieving and sustaining success in these complex ventures. According to Brandy and Davies (2014), there is an increasing awareness that using traditional project delivery approaches, which have been used in the past and especially when applied to the unique characteristics of today's mega projects, are inadequate.

Most agree that benchmarking is measuring the development of an organisation's project against other similar projects to ascertain the life-cycle duration and how much the project costs. Andersen and Pettersen (1996) imply that benchmarking requires a company to be humble enough to concede those other organisations and/or projects are better than they are. Subsequently, organisations should further accept that it would be prudent to use benchmarking data to learn how to equal or surpass others when undertaking their project. Accordingly, Lankford (2000) suggests that benchmarking is having a primary goal of using the knowledge and expertise of others to develop new and fresh ideas. Managers of organisations or projects with this type of insight would become evolutionary leaders to introduce a change in thinking or approach change that could alter the way that current mega complex LNG projects are managed

4.8.3 Development of Benchmarking Complex LNG Projects

Due to the unprecedented complexity of these projects, the current traditional benchmarking methods rooted in conventional project management techniques need to evolve significantly, which should be changed to allow for such complex possibilities currently not covered in any traditional aspect of project management techniques/processes. Goodwin and Goodwin (1999; p 416) state, "Many of the rules we encounter pertain to factor analysis, commonly used to estimate construct validity." As such, there is a necessity to look at the development of an instrument that validates a benchmarking construct for the enhancement of the life-cycle of mega/giga complex LNG projects. Goodwin and Goodwin (1999; p 410) state, "Validity is a multifaceted psychometric property." Benchmarking can be an invaluable tool for businesses to improve performance, but when done poorly, it can lead to significant wastage. Kozak (2004; p 5) states, "A poorly executed benchmarking exercise will result in a waste of financial and human resources, as well as time."

It is important to carefully plan and execute benchmarking exercises to ensure they yield meaningful insights and actionable results. This involves identifying clear objectives, selecting appropriate benchmarks, gathering accurate data, and analysing the findings effectively. Failure to do so can result in misinterpretation of data, misguided decisions, and ultimately, wasted resources.

The evolving landscape of mega/giga complex LNG projects necessitates a re-evaluation of traditional benchmarking methods to effectively address the complexities and challenges inherent in their development. Re-evaluation of traditional benchmarking methods for mega/giga LNG projects is essential to align with the evolving landscape, technological advancements, regulatory requirements, and societal expectations. Complex projects such as mega/giga LNG projects present multifaceted variables and unique challenges that conventional benchmarking approaches may struggle to address effectively. To address this, there is a need to adapt and innovate benchmarking techniques to encompass these complexities effectively. These projects involve a wide range of variables, including engineering, procurement, construction, logistics, regulatory compliance, multiple stakeholders, and environmental considerations, among others. To effectively benchmark such projects, innovative approaches are necessary.

Integrating advanced data analytics and artificial intelligence technologies into benchmarking processes represents a powerful approach to enhance the effectiveness of managing complex projects like mega/giga LNG developments. These technologies can handle vast amounts of data and identify patterns and insights that may not be apparent through traditional methods. Therefore, processes like machine learning techniques excel in identifying non-linear relationships and uncovering hidden correlations in data, enabling more accurate benchmarking comparisons. Consequently, there is a need to understand what machine learning is, in brief machine learning has five aspects to it as follows:

- ***Feature Learning***: provides algorithms which can automatically learn which features (variables) are most important and how they interact with each other. This is crucial for identifying complex relationships that might not be obvious from simple correlation analysis.
- ***Non-linear Relationships***: Many real-world relationships are not linear. Traditional statistical methods like linear regression struggle with non-linear data, but machine learning models such as decision trees, neural networks, and support vector machines can capture these non-linearities effectively.
- ***Dimensionality Reduction***: Techniques like Principal Component Analysis (PCA) or manifold learning can help reduce the dimensionality of data while preserving important non-linear relationships, making complex data more manageable and interpretable.
- ***Clustering and Pattern Recognition***: Machine learning algorithms can identify clusters and patterns within data that may reveal hidden correlations or groupings that are not apparent through traditional methods.
- ***Prediction and Generalisation***: By learning from data, machine learning models can make predictions about future observations or generalise patterns across larger datasets, allowing us to infer hidden correlations even in noisy or complex data.

Furthermore, embracing a more dynamic and flexible approach to benchmarking is essential. Rather than relying solely on periodic assessments, continuous benchmarking methodologies can provide real-time insights and allow for adjustments as project dynamics evolve.

Collaboration and knowledge-sharing among industry peers can also enhance benchmarking efforts. Participating in industry consortia or forums where organisations can exchange benchmarking data, and best practices can provide valuable insights into industry trends and performance benchmarks.

In essence, adapting traditional benchmarking methods to accommodate the complexities of modern projects requires a combination of advanced technologies, flexible methodologies, and collaborative approaches. By doing so, organisations can better position themselves to drive continuous improvement and innovation in their project management practices.

4.8.4 Definitions of Benchmarking LNG Projects

A defining feature of using benchmarking is the means of measuring a project against similar projects to reduce risks, especially when developing these complex LNG projects. Drawing from those with experience and involvement in developing these multiple mega complex LNG projects, there are no current acceptable means of benchmarking these types of projects due to the limited amount of mega complex LNG projects constructed. Subsequently, there is not a real universally accepted definition of the way to benchmark these projects as the projects differ between countries. Differencing factors such as remoteness, skilled resource availability, and infrastructure/environmental requirements vary and, accordingly, they have an impact as they are unique to the project locations.

When developing these projects, it is essential that organisations appreciate the aforementioned differencing factors and the local conditions early within the maturity of these complex projects. Consequently, having an appreciation of how organisations can gain by sharing their benchmarking data with other organisations and seeing the many benefits of doing so can provide the following:

- judicious project implementation throughout the project life-cycle, providing the phase durations use a banding duration frame to position the project based on location and complexity;
- accelerated infrastructure conception, i.e., benefits of pre-investments in the early development of the infrastructure;
- positional benefits of improved performances motivated by incentives to reduce costs and potentially improve management throughout the project life-cycle; and
- enhancement of unique management structures, which need to be applied to mega complex LNG projects to be modified from the structures used in conventional project developments.

Developing a universal benchmarking process for mega complex LNG projects holds significant potential for reducing cost and schedule overruns in future endeavours. To achieve this, it is crucial to understand the key differences between mega LNG projects and other types of oil and gas projects, as well as the specific nuances within the LNG sector itself.

4.9 Chapter Summary

Increased stakeholder involvement in the day-to-day decision-making process, including the selection of benchmarking data, reflects a growing recognition of the importance of collaboration and transparency in mega LNG projects. This trend can be attributed to several factors. Given the critical role of oil and gas energy security globally, particularly in regions like Australia, developing a standard, open approach for benchmarking oil and gas projects, including mega LNG projects, is of paramount importance. Overall, the development of a standard, open approach for benchmarking oil and gas projects, particularly mega LNG projects, is essential for advancing industry practices, ensuring energy security, and driving sustainable development in the oil and gas sector. Collaborative efforts among industry stakeholders, government agencies, and research institutions are critical in achieving this goal and maximising the benefits for all stakeholders involved.

The proposal for sharing data more freely between oil and gas organisations, particularly for mega LNG projects, holds tremendous potential for advancing industry practices and project outcomes. By establishing a benchmarking code of conduct specifically tailored to mega LNG projects, stakeholders can ensure that benchmarking efforts are conducted ethically, transparently, and in accordance with industry best practices. It also provides a structured methodology for managing these complex projects and the types of constructs that are necessary to be incorporated into organisations that are developing these multifaceted LNG projects. It includes the advancement of solid building blocks that are necessary to bring these mega complex LNG projects in on time and budget in future developments.

The evaluation methodology outlined in this chapter combines both quantitative and qualitative approaches to comprehensively review the strategies used to investigate the application of benchmarking and measures across the life-cycle phases of onshore LNG projects. By employing a combination of quantitative and qualitative approaches, the evaluation methodology ensures a comprehensive understanding of how benchmarking and measures are applied across the life-cycle phases of onshore LNG projects. This holistic approach allows for a nuanced analysis of both quantitative data and qualitative insights, leading to more robust findings and recommendations for improving benchmarking practices in the LNG industry.

The intention of this book is to advocate for the development of a more robust benchmarking process tailored specifically for mega/giga complex LNG projects. It argues that the current benchmarking data available is insufficient to address the unique challenges and complexities inherent in these projects. Instead, the book proposes the need for a comprehensive benchmarking process that involves gathering data from organisations that have experience with mega/giga LNG projects, both completed and ongoing.

By collecting robust benchmarking data from such organisations, the proposed benchmarking process aims to provide valuable insights into the performance, best practices, and lessons learnt throughout the entire life-cycle of mega/giga LNG projects. This includes benchmarking data related to project planning, design, construction, commissioning, operations, and decommissioning phases.

Overall, this book aims to improve decision-making processes, mitigate risks, and enhance the overall success of mega/giga LNG projects. Additionally, emphasising the importance of data collection from both completed and ongoing projects is crucial in acknowledging the dynamic nature of the LNG industry. This approach recognises that insights gained from past experiences can inform current and future decision-making processes. By analysing data from completed projects, stakeholders can identify patterns, trends, and lessons learnt, which can help in mitigating risks and improving project outcomes. Similarly, collecting data from ongoing projects allows for real-time adjustments and optimisations, ensuring that strategies remain relevant and effective amidst evolving market conditions and regulatory landscapes. This holistic approach to data collection supports continuous improvement and adaptation within the LNG industry.

Consequently, this ensures that the benchmarking process remains relevant and adaptable to evolving market conditions, technological advancements, and regulatory frameworks. This serves as a guide for stakeholders in the LNG industry to develop and implement a comprehensive benchmarking process tailored specifically for mega/giga complex LNG projects. By adopting such a process, organisations can leverage valuable insights and experiences from past projects to optimise the planning, execution, and management of future mega/giga LNG endeavours.

5 Mega Complex LNG Project Paradox

5.0 Defining a Paradox

A paradox is a statement or situation that appears to be contradictory or goes against common sense but may reveal an underlying truth or complexity. It often involves a contradiction or a conflict between two seemingly incompatible ideas or elements.

5.1 Chapter Introduction

The mega LNG project paradox refers to a recurring issue within multifaceted LNG projects where ambitious undertakings often face significant difficulties, resulting in their failure to meet initial objectives, such as budget, timeline, and performance targets. This paradox is not exclusive to LNG projects but is common in many mega infrastructure and industrial ventures.

Despite substantial investments in terms of time, resources, and finances, these projects experience massive cost overruns and large schedule creeps. Therefore, it is crucial to emphasise rigorous planning, realistic budgeting, effective risk management strategies, and clear communication throughout the project life-cycle. Regular monitoring and adjustment of plans based on actual progress can also help mitigate cost overruns and schedule creeps.

5.2 Mega Complex LNG Project Paradox

The paradoxes surrounding mega LNG projects challenge conventional wisdom and provoke deeper reflection on several fronts. These paradoxes underscore the complexity of large-scale LNG projects and highlight the need for structured, forward-thinking approaches to energy development and policy. They often highlight the limitations of human perception, reasoning, or language, revealing inherent contradictions within a particular concept or situation. In essence, the paradoxes associated with mega LNG projects highlight the complexities of balancing short-term energy needs with long-term sustainability goals. They underscore the limitations of human perception, reasoning, and language in navigating the multifaceted challenges posed by energy transitions and environmental stewardship

DOI: 10.1201/9781003688334-5

in a rapidly changing world. There are several factors that contribute to a Mega (complex) LNG projects paradox such as:

- **Complexity**: Mega LNG projects are inherently complex and multifaceted, due to their large scope, technical requirements, multiple stakeholders, and intricate logistics. Managing such complexities associated with these projects, which are developed in many different parts of the world, i.e., design houses and fabrication yards, can be overwhelming. This can lead to miscommunication, coordination issues, and, due to different time zones, difficulty in management making timely decisions. Managing the complexities associated with mega LNG projects requires a comprehensive and adaptive approach. By focusing on integrated project management, effective coordination, standardisation, risk mitigation, stakeholder engagement, technological innovation, and sustainability, project teams can navigate challenges more effectively and achieve successful project outcomes.
- **Uncertainty**: Mega LNG projects often operate in environments characterised by uncertainty, which poses significant challenges throughout their life-cycle. There are some key aspects of uncertainty in mega projects and strategies to manage them effectively. Factors such as changing market conditions, regulatory requirements, geopolitical risks, and technological advancements can introduce unpredictability, making it challenging to accurately plan and execute these projects on time and on budget. By proactively addressing uncertainty through comprehensive risk management, adaptive planning, stakeholder engagement, technological innovation, and compliance management, project teams can navigate unpredictable environments more effectively. Embracing uncertainty as an inherent aspect of large-scale projects and incorporating these strategies can enhance project resilience and increase the likelihood of achieving successful outcomes.
- **Stakeholder Management**: Mega LNG projects are inherently complex and involve a wide range of stakeholders, each with distinct interests, roles, and expectations. Managing these stakeholders effectively is crucial for the successful planning, execution, and operation of such projects. Balancing the interests and expectations of diverse stakeholders can be demanding, and conflicts may arise, leading to delays, disputes, and project disruptions. By effectively managing the diverse array of stakeholders involved in mega LNG projects through strategic engagement, transparent communication, proactive risk management, and sustainable practices, project developers can enhance project success, minimise conflicts, and contribute positively to local economies and environments.
- **Budget and Cost Management**: Cost overruns and budget constraints are common challenges faced by mega LNG projects due to their complexity, scale, and various contributing factors. Addressing these issues requires a comprehensive approach to project management, risk mitigation, and financial planning. Estimating the true costs of large-scale projects accurately can be difficult, and unforeseen expenses can arise during the Execute phase. Budget constraints may lead to compromises in quality, scope reductions, or inadequate risk mitigation measures.

- **Project Governance**: Effective project governance is crucial for the successful execution of mega LNG projects, which are characterised by their complexity, economy of scale, and diverse stakeholder involvement. However, governance structures may suffer from inefficiencies, lack of accountability, and inadequate risk management processes. Poor governance can contribute to misalignment of objectives, inadequate decision-making, and ineffective project controls. This also entails alignment of corporate governance and project governance. Consequently, effective project governance is a cornerstone of successful mega LNG projects. By establishing clear roles, strategic alignment, robust risk management, stakeholder engagement, and performance monitoring mechanisms, project teams can navigate complexities, mitigate risks, and achieve their objectives while delivering value to stakeholders and society at large.
- **Technical and Engineering Challenges**: Mega LNG projects push the boundaries of technology, engineering, and construction due to their scale and complexity. These technical complexities can introduce unforeseen technical challenges, requiring innovative solutions and expertise that may not be readily available. The success of mega LNG projects hinges on the integration of cutting-edge technologies, intricate engineering designs, and ambitious construction plans. By leveraging innovation, advanced engineering solutions, and strategic project management, stakeholders can overcome challenges, optimise performance, and deliver sustainable LNG solutions to meet global energy demands.

Addressing the mega (complex) LNG project paradox requires a comprehensive approach that focuses on proactive planning, effective risk management, robust project governance, stakeholder (corporate and project) engagement, and realistic estimation of costs and timelines prior to the Financial Investment Decision (FID) being approved. This comprehensive approach helps in navigating complexities and challenges that are inherent in mega LNG projects, fostering sustainable outcomes and maximising value creation. It is essential to prioritise transparency, collaboration, and continuous open and honest communication among all project stakeholders and the Joint Venture Partners (JVPs).

Furthermore, organisations and project teams should leverage lessons learned from previous mega projects, adopt best practices, and invest in building a strong project management capability to enhance the chances of success.

Recognising and addressing the complexities inherent in mega LNG projects is crucial for achieving successful outcomes. By implementing proactive measures, such as robust risk management strategies, streamlined project management practices, and thorough stakeholder engagement, organisations can better navigate the challenges posed by these projects. Additionally, fostering a culture of transparency, accountability, and continuous improvement can help mitigate risks and enhance overall project performance. Ultimately, by acknowledging and actively managing the mega LNG project paradox, organisations can increase the likelihood of delivering successful outcomes in this demanding and dynamic industry.

5.3 Cost Overruns Analysis

Analysing cost overruns in mega LNG projects is crucial for understanding why they often exceed initial estimates, despite thorough benchmarking during the project initiation phase. Several factors contribute to these overruns:

1. **Complexity**: Mega LNG projects are inherently complex, involving numerous stakeholders, intricate logistics, and extensive engineering requirements. Despite benchmarking efforts during the project initiation phase, unforeseen complexities can emerge during execution, leading to schedule delays and cost overruns.
2. **Geopolitical and Regulatory Challenges**: LNG projects often span multiple jurisdictions and are subject to various regulatory requirements and geopolitical factors. Changes in regulations, permits, or political climates can introduce delays and increase costs unexpectedly.
3. **Supply Chain Issues**: LNG projects rely on a global network of suppliers for materials, equipment, and skilled labour. Disruptions in the supply chain, such as shortages, price fluctuations, or quality issues, can impact project timelines and budgets.
4. **Technical Challenges**: LNG facilities require advanced technology and engineering expertise. Technical complexities, such as difficult terrain, adverse weather conditions, or novel design requirements, can lead to construction delays and cost escalations.
5. **Scope Creep**: Despite rigorous planning during the project initiation phase, scope creep can occur as stakeholders identify additional requirements or changes in project specifications. Managing scope changes effectively is crucial in mega LNG projects to maintain project schedules and prevent cost escalation. Consequently, effective scope change management is integral to the success of mega LNG projects. By adhering to structured processes, engaging stakeholders proactively, and maintaining alignment with project objectives, teams can navigate scope changes while minimising disruptions and optimising project outcomes.
6. **Contractual Disputes**: Disputes between project owners, contractors, and subcontractors over scope, pricing, or project specifications can result in legal battles and project delays. Resolving contractual disputes often requires time and resources, contributing to cost overruns.
7. **Currency Fluctuations and Economic Factors**: Mega LNG projects are typically long-term investments spanning several years. Exchange rate fluctuations, inflation, and economic downturns can significantly impact project costs, especially for projects with multinational stakeholders.
8. **Environmental and Social Factors**: Increasing awareness of environmental and social impacts necessitates compliance with stringent regulations and community engagement efforts. Failure to address these factors adequately can lead to project delays and increased costs.

Addressing these challenges requires robust risk management strategies, effective communication among stakeholders, and agile project management practices.

By proactively identifying potential risks and implementing mitigation measures, project teams can minimise the likelihood of cost overruns and improve project outcomes. Aritua et al. (2009; p 74) state, "Numerous authors have shown that management of multi-projects is not simply an aggregate of single project efforts and, as such, require unique approaches, techniques, and tools."

The objective is to identify why, even with benchmarking, mega/giga LNG projects often overrun on schedule and budget due to a variety of complex and interrelated factors. Love et al. (2013) make a crucial distinction between cost overruns and cost escalations, the latter being the anticipated growth of a budget due to factors like inflation, etc. Benchmarking provides valuable insights and comparisons, but it cannot entirely predict or mitigate the myriad challenges that mega and giga LNG projects face. The interplay of these factors often means that projects will encounter unforeseen issues that can lead to schedule overruns and budget increases. Addressing these challenges requires a dynamic approach to project management, risk assessment, and flexibility to adapt to the changing circumstances. Cost overruns in mega LNG projects not only represent the deviation from the original budget and scope but also highlight the challenges associated with quantifiable benchmarking. The complexity of these projects can make it difficult to define the scope of work accurately at the outset, leading to uncertainties that may not be fully captured in benchmarking efforts. Aritua et al. (2009; p 72) state, ". . . the need for new approaches, processes, and techniques suitable for multi-project management is obvious. This, however, requires a change in mindset . . .". Love et al. (2013) suggest that cost overruns are more likely in larger projects than smaller ones due to their complexity. Stingl and Geraldi (2017) point out that when there is an escalation of commitment this becomes a typical problem in large projects and, as such, they continue although they have significant cost overruns and extreme delays indicating project failure. Therefore, project managers and organisations need to understand how these mega LNG projects are being benchmarked and what matrix is being used, including whether factors such as project location, remoteness, etc., were considered. The outcome from this process assists in providing a tangible means for conducting benchmarking in future mega (complex) LNG projects worldwide. By gaining a comprehensive understanding of how mega LNG projects are benchmarked and the factors considered in the process, project managers and organisations can improve future project planning and execution, enhance risk management practices, and ultimately achieve better project outcomes. By leveraging insights from benchmarking exercises and applying them to future projects, project managers can reduce the likelihood of schedule and cost overruns in large complex projects like mega LNG developments.

5.4 Why Conduct a Case Study

A case study is a research method that involves an in-depth and detailed examination of a specific individual, group, event, organisation, or phenomenon. Case studies are used for various purposes in different fields, including psychology, sociology, business, medicine, education, oil and gas projects, research, and more. Some reasons why case studies are valuable as they provide:

- **In-Depth Understanding**: Case studies allow researchers to gain a deep and comprehensive understanding of a particular subject. By focusing on a single case, researchers can explore nuances, complexities, and context that may be overlooked in broader studies.
- **Contextual Analysis**: Case studies enable the examination of a subject within its real-world context. This can provide insights into how various factors interact and influence outcomes.
- **Theory Development**: Case studies can help in the development or refinement of theories. Researchers can use the data collected from a case study to build or modify theories and hypotheses.
- **Problem Solving**: In fields like business, case studies are often used to analyse and solve real-world problems. Organisations can learn from past experiences to make better decisions and avoid mistakes.
- **Exemplification**: Case studies can be used to illustrate key concepts or principles. They provide concrete examples that can enhance teaching and learning.
- **Pilot Research**: Case studies can be a starting point for research in an area that is not well understood. They can help identify research questions and hypotheses for further investigation.
- **Rich Data**: Case studies typically involve the collection of various types of data, such as interviews, observations, documents, and artifacts. This rich data can provide a holistic view of the subject.
- **Longitudinal Studies**: Some case studies are conducted over an extended period, allowing researchers to track changes and developments over time.
- **Uniqueness**: Some phenomena or situations are unique or rare, making case studies the only practical research method. For example, the uniqueness of developing a mega LNG project, such as the Gorgon LNG project, which involved developing a class A nature reserve.
- **Policy and Practice Evaluation**: Case studies can be used to evaluate the effectiveness of policies, interventions, or practices in specific settings.
- **Interdisciplinary Research**: Case studies can bridge gaps between different disciplines by examining issues from multiple angles.
- **Real-Life Application**: The findings from case studies can have direct applications in real-life settings, making them valuable for practitioners and policymakers.

However, it is important to note that while case studies offer many advantages, they also have limitations. They may not be representative of larger populations, and generalising findings from a single case can be challenging. Researchers must carefully consider the scope and purpose of their study when choosing the case study method.

The case studies provide a means of generating hypotheses rather than consistent information about the overall project outcome. Giezen et al. (2014; p 1004) state, "A case study is an empirical inquiry that investigates a contemporary phenomenon within its real-life context, especially when the boundaries between phenomenon and context are not clearly evident."

Accordingly, the case studies are useful for providing hypotheses for the development of a full research process. Miles and Huberman (1994; p 1) state, "With qualitative data one can preserve chronological flow, see precisely which events lead to which consequences, and derive fruitful explanations." Furthermore, Miles and Huberman (1994; p 1) state, ". . . explanations flow from an account of how differing structures produced the events we observed."

5.5 Case Study

Navigating the Political Landscape: Hypothetical Case Study of LNG Projects in Australia

Overview: Australia, renowned for its vast natural resources, has witnessed a surge in liquefied natural gas (LNG) projects over the past decade. This theoretical case study delves into the intricate political dynamics surrounding the development and regulation of LNG projects in Australia, exploring the interplay between government policies, industry interests, environmental concerns, and Indigenous rights.

By providing a structured approach to mega LNG project development, this framework can help project stakeholders navigate the complexities involved and achieve successful outcomes. It is essential for the framework to be flexible and adaptable to different project contexts and evolving industry trends. Additionally, validation and refinement of the framework through real-world applications and feedback from industry experts would enhance its effectiveness and relevance.

LNG projects involve numerous complexities, including technical, financial, regulatory, and environmental considerations. Having a comprehensive framework can streamline the development process and mitigate risks.

Key components of such a framework include:

1. **Project Planning and Feasibility Analysis**: Assessing the technical feasibility, market demand, and economic viability of the project;
2. **Regulatory and Permitting Processes**: Navigating the regulatory landscape, obtaining permits, and ensuring compliance with environmental and safety regulations;
3. **Stakeholder Engagement**: Engaging with stakeholders, including local communities, government agencies, environmental groups, and investors, to address concerns and gain support for the project;
4. **Risk Management**: Identifying potential risks and developing strategies to mitigate them, such as through risk assessment, contingency planning, and insurance;

5. **Supply Chain Management**: Managing the supply chain, including procurement of equipment and materials, logistics, and transportation of LNG;
6. **Project Financing**: Securing financing for the project, including debt, equity, and government incentives, and managing financial risks;
7. **Construction and Commissioning**: Planning and executing the construction phase, ensuring quality control, safety, and adherence to schedule and budget;
8. **Operations and Maintenance**: Establishing procedures for the operation and maintenance of the LNG facility, including monitoring performance, maintenance schedules, and emergency response plans;
9. **Sustainability and Social Responsibility**: Integrating sustainability principles and social responsibility practices into the project, including environmental impact assessments, community engagement, and corporate social responsibility initiatives; and
10. **Technology and Innovation**: Incorporating innovative technologies and best practices to optimise efficiency, reduce costs, and minimise environmental impact.

Case Study: GreenWave LNG Terminal

Background:

GreenWave Energy, a multinational energy corporation, plans to develop the GreenWave LNG Terminal, a mega LNG project located on the coast of a developing country. The terminal aims to liquefy and export natural gas to global markets, capitalising on abundant natural gas reserves in the region.

Challenges:

1. **Regulatory Hurdles**: The project faces regulatory challenges related to environmental impact assessments, land acquisition, and compliance with local laws and regulations;
2. **Stakeholder Opposition**: Local communities, environmental groups, and indigenous populations oppose the project due to concerns about environmental degradation, displacement, and social disruption;
3. **Financing Complexities**: Securing financing for the project proves challenging due to geopolitical risks, fluctuating commodity prices, and the uncertain regulatory environment;
4. **Infrastructure Development**: Building the necessary infrastructure, including LNG liquefaction plants, storage tanks, pipelines, and marine facilities, requires significant investment and expertise; and
5. **Operational Risks**: Operating the LNG terminal entails risks such as technical failures, natural disasters, security threats, and market volatility.

Application of the Framework:

1. **Project Planning and Feasibility Analysis**:
    - conduct a comprehensive feasibility study to assess the technical, economic, and market feasibility of the project; and
    - identify potential risks and mitigation strategies, including regulatory compliance, stakeholder engagement, and financial risk management.
2. **Regulatory and Permitting Processes**:
    - engage with government agencies, regulatory bodies, and local communities to obtain necessary permits and approvals; and
    - conduct environmental impact assessments and develop mitigation plans to address concerns raised by stakeholders.
3. **Stakeholder Engagement**:
    - establish a stakeholder engagement plan to communicate with local communities, indigenous groups, environmental organisations, and government officials; and
    - address concerns and grievances raised by stakeholders through transparency, consultation, and participation in decision-making processes.
4. **Risk Management**:
    - develop a comprehensive risk management plan to identify, assess, and mitigate risks associated with project development and operation; and
    - implement risk mitigation measures, such as insurance, contingency planning, and diversification of financing sources.
5. **Supply Chain Management**:
    - procure equipment, materials, and services from reliable suppliers and contractors, ensuring quality, cost-effectiveness, and timely delivery; and
    - manage the supply chain efficiently to minimise delays, disruptions, and cost overruns.
6. **Project Financing**:
    - explore various financing options, including project finance, export credit agencies, multilateral development banks, and private equity; and
    - mitigate financial risks through hedging strategies, revenue optimisation, and debt restructuring.
7. **Construction and Commissioning**:
    - execute the construction phase according to schedule, budget, and quality standards, employing experienced contractors and project management professionals; and

- commission the LNG terminal systematically, conducting thorough testing, training, and safety audits before commercial operation.

8. **Operations and Maintenance**:
    - establish robust procedures and protocols for the safe and efficient operation of the LNG terminal, including maintenance schedules, emergency response plans, and performance monitoring; and
    - invest in employee training, technology upgrades, and preventive maintenance to maximise uptime and minimise downtime.
9. **Sustainability and Social Responsibility**:
    - integrate sustainability principles into project design and operation, implementing measures to minimise environmental impact, conserve natural resources, and support local communities; and
    - implement social responsibility programs, such as community development initiatives, education, healthcare, and job training, to enhance social welfare and mitigate social risks.
10. **Technology and Innovation**:
    - embrace technological innovations and best practices to optimise LNG production, storage, transportation, and regasification processes; and
    - invest in research and development to enhance energy efficiency, reduce emissions, and improve safety standards.

Outcomes:

- by applying the framework systematically, GreenWave Energy successfully navigates the complexities of developing the GreenWave LNG Terminal;
- the project obtains necessary permits and approvals, addresses stakeholder concerns, and secures financing from a consortium of international lenders and investors;
- the GreenWave LNG Terminal becomes operational on schedule, exporting LNG to global markets and contributing to economic growth, energy security, and environmental sustainability in the region; and
- GreenWave Energy establishes itself as a leader in the LNG industry, renowned for its commitment to excellence, innovation, and corporate social responsibility.

In summary, the hypothetical case study illustrates how the framework for mega LNG projects provides a guideline for overcoming challenges and achieving success in project development and operation. It underscores the importance of systematic planning, stakeholder engagement, risk management, benchmarking, and sustainability to ensure the viability of mega LNG projects in an increasingly complex and competitive global market.

Key Events:

The author's research has established the need for a new type of outline and provided a benchmarking framework for mega LNG projects that provides a guideline for developing these projects; and

This framework provides a structure and guidance for understanding complex concepts or phenomena, making them invaluable tools for researchers and practitioners alike.

Key Themes:

1. **Resource Governance**: The case study examines the governance frameworks and regulatory mechanisms governing the extraction, processing, and export of LNG, exploring issues of transparency, accountability, and regulatory effectiveness.
2. **Stakeholder Engagement**: The complex network of stakeholders, including government agencies, industry players, environmental groups, communities, and international partners, underscores the importance of inclusive decision-making processes and effective stakeholder engagement strategies.
3. **Sustainable Development**: Balancing economic development objectives with environmental sustainability goals and social equity considerations presents a formidable challenge for policymakers and industry stakeholders, highlighting the need for integrated approaches to sustainable development planning and implementation.
4. **Risk Management**: The inherent risks associated with LNG projects, including market volatility, environmental impacts, social conflicts, and regulatory uncertainties, necessitate robust risk management strategies and contingency planning to mitigate adverse outcomes and enhance project resilience.
5. **Benchmarking**: The ever-increasing need for organisations to not just benchmark their projects at the start, Assess/initiate phase, of the project but to revisit the benchmarking as the project matures through each phase. Ensure that due diligence is undertaken on the benchmarking prior to entering the risk gates at the end of each phase of the life-cycle.

Analysis:

Through a hypothetical exploration of LNG projects in Australia, researchers can analyse the complex interplay of political, economic, social, and environmental factors shaping the development and governance of natural resource projects in a democratic context. By examining the challenges and opportunities inherent in LNG development, stakeholders can identify policy options, best practices, and lessons learned to inform decision-making processes and enhance the sustainability and resilience of energy development initiatives in Australia and beyond.

5.6 Chapter Summary

The author's observations gleaned from the case studies of mega LNG projects offer invaluable insights that contribute to the analysis of these multifaceted endeavours. Mega LNG projects are characterised by their complexity, involving numerous interconnected components, stakeholders, and challenges. Analysing such projects requires a comprehensive understanding of various factors, including technical requirements, regulatory frameworks, environmental considerations, supply chain dynamics, geopolitical factors, and financial constraints.

The foregoing demonstrates that these mega/giga LNG projects are multidimensional projects due to their complexity. These observations have demonstrated that it should be a prerequisite, at the risk stage gates, to spend the time and effort to revisit the benchmarking durations along with the scope of works and governance process.

The recommendation is that organisations should not merely accept the data provided by a single benchmarking organisations (house) but should also compare their data (provided) with other data from reputable benchmarking organisations (houses) as a normalisation process. The argument is that organisation stakeholders would not agree to the need to revisit the benchmarking, as they have used a reputable benchmarking organisation. Therefore, it would be challenging to convince stakeholders to fund this proposal. This could demonstrate short-sightedness by management and stakeholders of such organisations.

The flat management structures used on smaller LNG projects are too loose for these mega (complex) LNG projects. The benchmarking process should be relevant to projects in the location in which they are being developed, and, therefore, need to be governed with a wider spectrum of data. This is mainly due to the remoteness of locations, such as Australia, where there is limited local skilled management and resources, and distances from fabrication yards and design organisations where silos can be created, etc.

The author's observations show that this investigation is of a truly exploratory nature for developing these mega/giga (complex) LNG projects. It does not appear that this type of research in this area of the oil and gas industry for mega/giga LNG projects has previously been undertaken.

6 Life-Cycle Model

6.0 Overview

Understanding the life-cycle model is crucial in project management as it provides a structured approach to managing the stages a project goes through from its inception to its conclusion. By following the life-cycle model, project managers can systematically manage resources, timelines, and stakeholder expectations throughout the project life-cycle. It provides a framework for organising and prioritising tasks, reducing risks, and maximising the chances of project success. Each phase of the life-cycle plays a crucial role in achieving the overall goal of delivering a successful project. These phases have their own set of activities and objectives, contributing to the overall goal of delivering a successful project. They provide a framework for understanding and managing the activities, tasks, and deliverables for each phase of the project. The choice of the life-cycle model depends on various factors, including the nature of the project, stakeholder preferences, and the level of uncertainty and complexity in requirements. Ultimately, the life-cycle model provides a structured approach to manage the various phases and milestones of a project, ensuring efficient and effective development and delivery of the project.

Other life-cycle models, such as iterative or agile approaches, involve cycles of development, testing, and feedback to continuously improve the project. These models emphasise flexibility, collaboration, and adaptability to changing requirements.

6.1 Chapter Introduction

This chapter discusses the development of the life-cycle model for a hypothetical LNG Performance Measurement System (PMS). Hussein et al. (2022; p 1) state "the first hybrid life cycle sustainability assessment (LCSA) of liquefied natural gas and evaluate its performance from the natural gas extraction stage to LNG regasification after delivery through maritime transport carriers." The main sections of this chapter demonstrate how the life-cycle model findings contribute to the author's investigation and how these form part of the performance measurement system. Schwarz (1995; p 155) states, "When one looks at textbooks of survey methodology, one comes away with the impression that question comprehension is mainly an issue of choosing the right words."

DOI: 10.1201/9781003688334-6

The issues of management resources movement on mega/giga (complex) LNG projects is crucial due to the scale, complexity, and geographic spread of these projects. Identifying these resource management issues requires a holistic approach, involving collaboration between project teams, stakeholders, and external partners. It identifies the approach the project team takes to align the KPIs, and CIs associated with the movement of personnel during each phase of the project life-cycle. Establishing these metrics with the project's objectives is crucial for tracking progress and ensuring that the project remains on track. As a result, the need to frequently review and monitor the individual and project KPIs are a management obligation/responsibility to ensure that they are relevant to the success of the team and project outcome. Schonberger (2008; p 6) states, "Mid-range and higher goals and KPIs get monitored and acted on (assigning responsibilities) too frequently. Monthly is much too often to assess complex programs and initiatives." Accordingly, there needs to be a process and a balance regarding when these are reviewed, noting that most KPIs set for individuals are really outside the sphere of influence of the project management team. Frequent review and monitoring of KPIs are essential management responsibilities that help ensure the success of both the team and the project. Regularly reviewing and monitoring both individual and project-level KPIs is vital for ensuring that they remain relevant and aligned with the team's and project's objectives. This ongoing assessment allows management to identify any deviations from the desired outcomes early on, enabling them to take corrective actions promptly. Additionally, the management of visibility to the team is necessary for the use of CSFs, and they can capture the tacit knowledge that individuals bring to the team from other projects.

6.2 Performance Indicators

The life-cycle model reviews the Key Performance Indicators (KPIs), Critical Success Factors (CSFs), and Core Indicators (CIs) and how they need to be communicated to the project team throughout the life-cycle development of these multifaceted and massive LNG projects.

In the context of such LNG projects, the life-cycle model is particularly critical due to the complex and difficult nature of these projects. LNG projects involve significant investments, extensive engineering, regulatory compliance, and coordination among multiple stakeholders. Applying the life-cycle model to LNG projects helps manage the complexities and ensure that all aspects of the project are planned, executed, and reviewed effectively. This structured approach is crucial for delivering LNG projects on time, within budget, and to the required quality standards.

The importance or awareness of how-when-who of developing key items throughout each phase of the project life-cycle is critical for effective project management, particularly in complex and high-stakes projects like LNG. These items should be transparent to align the team competencies to evaluate the performance and undertake the tasks. This awareness ensures that all aspects of the project are managed effectively and that the project progresses smoothly towards the completion.

Marrewijk et al. (2008; p 592) state, ". . . megaprojects involve multiple competencies, each of which will be characterised by their specific rationalities, such that talk, decisions and actions will not necessarily be aligned with each other." Subsequently, these KPIs, CSFs, and CIs, depending on the size of the project, can contain many varied factors. These elements can be numerous and can become a very advantageous means of successfully managing and measuring the project and individuals' success, especially when managing multidimensional mega/giga (complex) LNG projects. Toor and Ogunlana (2010) identify that success factors are those that contribute to the success of a project by measuring against cost, time, and quality to show how and when accomplishments are measured with success factors such as CSFs. Furthermore, Ram et al. (2013; p 159) state ". . . a factor can only be termed a CSF if attending to this factor in a satisfactory manner results in performance improvements." Therefore, identifying whether the influence of a CSF is critical with the inherent complexity of mega/giga LNG projects can be a key factor in increasing the project's performance. Sobieraj and Metelski (2021; p 2) state, "Critical success factors are important because they help project managers to adequately communicate between different project stakeholders, monitor and control the scope, change and risk, identify and prioritize goals and objectives, identify and allocate resources, and remove bottlenecks in knowledge exchange flows."

6.2.1 Key Performance Indicators

These are measurable metrics that gauge the performance of specific aspects of the project. KPIs can vary depending on the project's objectives but may include metrics such as cost performance, schedule adherence, safety records, and environmental impact. Throughout the project life-cycle, KPIs should be continuously monitored, analysed, and communicated to the project team to track progress and ensure alignment with project goals. As the project team and management change out in most phases, especially the Assess, Select, and Develop phases, it is recommended that the KPIs are aligned to new team members' competencies as they come on board. This ensures that the introduction of any additional KPIs is appropriate to the skills of the new personnel and is linked with the existing project objectives of how they are measured to achieve project success. Kerzner (2014) suggests that KPIs measure the quality of the process to achieve the end results and, as such, are internal measures that are reviewed on a periodic basis throughout the project life-cycle. These measures are based on schedule, cost, and quality KPIs, which are agreed with stakeholders, for project success and customer satisfaction.

The use of KPIs in mega/giga (complex) LNG projects is commonly aligned with the project Critical Success Factors (CSFs). Sobieraj and Metelski (2021; p 3) state, "CSFs can be tracked through key performance indicators (KPIs), by means of which one can adequately assess the management of all project elements." Therefore, it is important that, at each phase, the KPIs (project and individual) are revisited frequently, as a continuing process of best practices, to ensure they remain aligned with the original outcome for the project success. Love et al. (2009) suggest, even with a transient period, knowledge cannot be readily passed on from one individual to another and that new project members cannot acquire sufficiently detailed project knowledge immediately after joining a project or phase. Kerzner (2014; p 44) states ". . . best practices come from knowledge transfer and can be discovered anywhere within or outside of your organisation." Consequently, it is essential that companies setup a best practice process or documentation to capture all the best practices from knowledge transfer (tacit information) from existing and new project team members, which can be part of developing the Critical Success Factors (CSFs) early on in the initial project development. Kerzner (2014; p 135) states, "The CSFs and KPIs are the drivers for project management to become a strategic competency and a competitive advantage."

6.2.2 Critical Success Factors

These are the essential areas or activities that must be performed effectively to ensure the success of the project. CSFs are often broader in scope than KPIs and focus on strategic aspects that are crucial for achieving project objectives. CSFs are business objectives that are developed in the Assess (initiate) phase at the project kick-off meeting with senior management, JVPs, and the project team. Kerzner (2018; p 32) states, "CSFs identify those factors necessary to meet the desired deliverables of customers." This establishes the initial scope, timeframe, and budget requirements. Kerzner (2014) points out that the use of CSFs is to map out the overall business objective of the project. With every project, there are different CSFs as they do differ from project to project; therefore, CSFs are potentially not transferable between projects. Toor and Ogunlana (2010) suggest that success factors do have some common characteristics between projects, such as limited budgets, difficult and complex schedules; however, they do vary from project to project. How then does the application of CSFs benefit the development of mega/giga (complex) LNG projects? Mavi and Standing (2018; p 752) state, "Emphasizing the CSFs enhances the capability of organisations to perform a project through its full execution." The CSFs are potentially a mechanism used by organisations to ascertain how they actually select the project team members, which is all part of controlling the project performance effectively. According to Mavi and Standing (2018), reviewing the project management and organisational CFSs is an essential measurement for assessing a project's success. Toor and Ogunlana (2010) suggest that CSFs can be used for the selection of the project team members by utilising the selection of their strategic standing on the project. It should also be recognised that the application of CSFs could provide an insight into the effective utilisation of limited resources, such as time and manpower availability and budgetary needs of the project. Toor and Ogunlana (2010; p 153) states, "Identification of CSFs enables the effective allocation of limited resources such as manpower and budget for the project." Kerzner (2014; p 482) states, "Critical Success Factor (CSF): This measures changes in the output of the project resulting from implementation of best practices."

6.2.3 Net Present Value

This is a critical financial metric used to assess the viability and profitability of a project, particularly in the context of investment decisions. NPV represents the difference between the present value of cash inflows and outflows over the project's life-cycle, discounted at a specified rate of return. It helps determine whether a project will generate positive or negative returns and by how much.

NPV serves as a key criterion for passing through the risk gates, especially the final risk-gate, which provides the FID approval (or not) to move into the Execute phase. Magni and Marchioni (2020; p 1) state, "In capital asset projects, economic profitability may be measured with absolute metrics, such as the net present value (NPV), expressing value increase in monetary units, or relative metrics, expressing rates of return or profitability indices which aim at identifying a project's financial efficiency." The NPV is a technique of planning and evaluating cost-plus time and cost-plus risk, noting there is a specific discount rate to cash flow of the project that provides an absolute measure. Priemus and van Wee (2013; p 43) state, "NPV takes expected cash flows (revenues and costs) in each period and combines them into a single present value using a discount rate that reflects both the price of time and the price of risk." Therefore, a positive NPV (NPV>0) for example would indicate to the JVPs that it is actually viable to continue with the project. Akampurira (2014; p 31) states, "The NPV which is positive implies the project is acceptable . . . negative implies the project is not acceptable." To provide a better appreciation of the NPV, Magni and Marchioni (2020; p 3) state, "A project creates value (i.e., it is worth undertaking) if and only if NPV>0." As an observation, any value can be used to pass a project through risk gates; as such, it must be backed up by robust data to ensure that senior management does not force a positive NPV to ensure the project continues. Priemus and van Wee (2013; p 43) state, "Caveats to NPV techniques are well known. The biggest drawback in early planning stages is that it is difficult if not impossible to fully specify the distributions of various cash flow elements, and hence expectations, over the life of the mega project."

This overview of the NPV highlights the potential risk in way organisations perceive or use the NPV to pass through the final risk gate during the attempt to achieve the FID approval and move into the Execute phase.

It is suggested that more investigation is carried out on how tacit information, KPIs, CSFs, and NPV calculations could be framed to ensure that there is no distortion that would or could be biased to any benchmarking data available and to the project's overall success.

NPV plays a crucial role in the decision-making process, especially when it comes to assessing the financial viability of projects and determining whether they should proceed to the Execute phase. By considering NPV alongside other relevant factors, stakeholders can make well-informed decisions that maximise value and mitigate risks.

6.3 LNG Life-Cycle Performance Measurement Model

Integrating both conceptual modelling and empirical modelling is crucial for comprehensive analysis and evaluation, especially in complex scenarios like case studies involving major components. Ray et al. (2015; p 78) state, "By far the most commonly used and also the most easily understood measure of performance is productivity."

By combining conceptual and empirical modelling approaches, analysts can achieve a more comprehensive understanding of the case study's major components and their relationships. Conceptual models provide a theoretical framework for analysis, while empirical models validate these theories using real-world data. This iterative process of modelling, testing, and refining hypotheses helps ensure that the findings are robust and actionable, ultimately leading to more informed decision-making.

Throughout the earlier chapters, it has been demonstrated how the Core Indicators (CIs) and Key Performance Indicators (KPIs) are linked together to provide a comprehensive measurement process for the life-cycle phases of mega/giga (complex) LNG projects. This measurement process must also include the Earned Value Management System (EVMS) described in section 2.4 General Construction Performance Measurements. According to Fleming and Koppelman (2010), the Earned Value Management concept requires a baseline plan, known as the planned value, using project scheduling systems, where performance is measured against the planned value.

Figure 6.1 illustrates a life-cycle model for these mega (complex) LNG projects, which provides the roadmap of how these multifaceted projects are developed and the interaction with CIs and stakeholders.

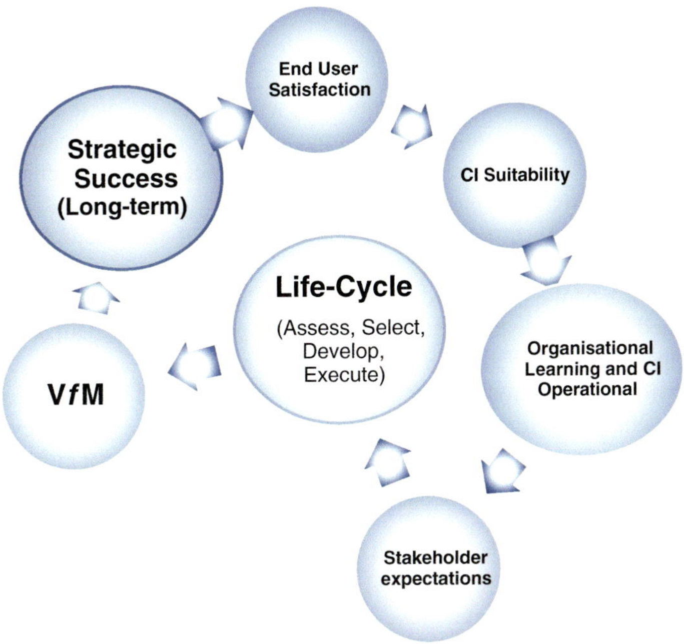

Figure 6.1 Roadmap for Development of a Life-Cycle Model

Source: Author's Research

6.4 Efficiency/Performance Measures

According to Iyer and Banerjee (2015), the use of Earned Value and Critical Chain Project Management methods is known to improve project performance/efficiencies, as described in the Project Management Body of Knowledge (PMPOK). Whereas Egan (1998) mentions that there are efficiencies in projects when teams are using Value Management and Benchmarking, which have been identified as the two major performance improvement methods that can significantly enhance efficiencies in project development.

Even though value management attempts to improve the project outputs (products, services, or outcomes), there is still a need for organisations to understand what they are trying to measure and why. Arndt and Edrington (2011; p 1) state, “Benchmarks are typically established through the process of identifying successful business practices through performance measurement.” As such, the overall performance measures need to be agreed upon when working within the boundaries of a project. This includes benchmarking efficiencies that determine the organisation’s objectives to be aligned with the best industry practices available. This leads to exceptional performance through the application of these best practices when developing projects. Baek and Lee (2009; p 265) state, “Efficiency analysis is performed not only to estimate the current level of efficiency, but also to provide information on how to remove inefficiency, that is, to obtain benchmarking information.”

6.5 Different Types of LNG Projects

The author’s observations suggest that to define LNG projects requires an understanding that they vary in size and complexity, and understanding this variability is essential for effective planning and development. The concept of economy of scale is particularly pertinent in the context of LNG projects, as it influences various aspects of project design, construction, and operation. Love et al. (2015) suggest that there is a need to forge a mutual understanding of what complexity means within a project life-cycle.

The author’s analysis of the development of these complex LNG projects has been divided into four distinct categories for LNG projects, i.e., small conventional LNG projects, large conventional LNG projects, large complex LNG projects, and exceptionally large/extraordinarily complex LNG projects. This provides specific benchmark constructs for organisations for each phase and overall life-cycles. The author’s categories, are divided into four sectors based on the economy of scale and complexity, and are proposed:

- **Small Conventional LNG Project**
  Not complex + small + under US $1 billion. These typically involve lower production capacities and serve niche markets or specific geographic regions.
- **Large Conventional LNG Projects**
  Not complex + large + over US $1 billion under US $10 billion. These involve significant production capacities and serve broader markets, often involving international trade.

- **Large Complex LNG Projects**
  Complex + very large + over US $10 billion and under US $20 billion = mega project
  (Example: Pluto LNG Project (CAPEX) ~US $13 billion (large = single train.)
  They are characterised by additional technical, logistical, or regulatory complexities beyond typical large-scale projects.
- **Very Large/Very Complex LNG Projects**
  Extraordinarily complex + very large + over US $20 billion = giga project. (As mentioned in Chapter 1). Example: Gorgon LNG Project (CAPEX) ~US $54 billion (economy of scale = three trains.)

These "Very Large/Very Complex LNG Projects" represent the pinnacle of scale and complexity within the LNG industry.

These projects often push the boundaries of technology, engineering, and project management capabilities.

The "Very Large/Very Complex LNG projects" described as giga projects (as defined in Chapter 1), are the projects that are so complex and multifaceted in nature, the author suggests that they have a CAPEX value of ~US $20 billion or greater.

The author's observations suggest that a project which would fall under the "Very Large/Very Complex" LNG projects would be the complex Chevron Australia Gorgon LNG project, which was ~US $54 billion value that was developed on class 'A' reserves on Barrow Island, in the northwest of Australia. Accordingly, the author suggests that this multifaceted LNG project would also be categorised as a giga project due to its size and complexity. This complex LNG project contained several stand-alone mega projects. Previously, Merrow (2011) and other authors have defined similar mega projects with a CAPEX benchmark of US $1 billion and above. This project comprised several separate sections containing mega projects, each with a value of around ~US $10 billion. Hence, the requirements to redefine the status of LNG projects and categorise them by the measures noted above.

By classifying LNG projects into these four distinct categories, stakeholders can better understand the varying scale and complexity of projects within the industry, enabling more targeted planning, investment, and risk management strategies.

There are several variations of the life-cycle model, but a common example is the waterfall model, which consists of sequential stages as demonstrated below:

- **Requirements gathering**: In the gathering phase of a project, the goal is to clearly define what the project needs to achieve and ensure that these needs align with the stakeholders' expectations. The project team identifies the documents, which also include the stakeholders' requirements;
- **Design**: In the design phase of a project, the focus shifts from understanding what needs to be done (requirements gathering) to defining how it will be done. This involves creating a detailed plan that outlines the structure, components,

and interactions needed to fulfil the project requirements. By carefully planning and documenting the design, it sets a clear path for the development and execution of the project, helping to ensure that it meets the defined requirements and is delivered successfully;

- **Development**: In the development phase of a project, the focus is on translating the detailed design into a functional product or system. This phase involves coding, building, and assembling the components as specified in the design. By carefully managing each step of the development phase, you ensure that the project is built according to the design specifications and is ready for deployment and use;
- **Testing**: The testing phase is crucial in ensuring that the project meets the defined requirements and functions as expected. This phase involves systematically checking the implementation against the requirements to identify and resolve any issues before the final release. Checking against the requirements to ensure it functions correctly and meets the stakeholders' expectations is a crucial process; and
- **Deployment**: The deployment phase is the final step where the project is released or delivered to its intended users or stakeholders. This phase involves transitioning from development and testing to operational use.

Accordingly, as mega LNG projects become more complex, increasingly large, technically challenging, graphically diverse, more multifaceted, and expensive, conventional management techniques and approaches may become inadequate for effectively managing them. Haidar and Ellis (2010; p 9) state, "classical planning and management is not suitable for megaprojects and can result in bureaucratic, inflexible, and powerless organisations."

Subsequently, there is a further need to modify the way we approach the performance measurement systems required to manage these complex LNG projects, such as KPIs, BSCs, and EVMS, which can vary from managing each of the groups identified above.

By providing guidelines and benchmarks for organisations relating to the types of LNG projects they intend to develop, this consequently enhances clarity, transparency, and decision-making effectiveness throughout the project life-cycle. By leveraging these definitions and categorisations, organisations can optimise their project development efforts, mitigate risks, and maximise the likelihood of achieving successful outcomes in the dynamic and complex LNG industry.

The author's observation is that it is realistic to suggest that the difference between complex LNG projects is quite easy to define and benchmark. Accordingly, it is as straightforward as the reservoir size, which then determines the CAPEX values of the project associated with the economy of scale to produce the level of mtpa as the benchmark. While reservoir size and production capacity (measured in million tonnes per annum, mtpa) are important factors in defining the complexity and scale of LNG projects, it is important to note that the differences between complex LNG projects are not solely determined by these factors alone. While reservoir size can indeed influence project complexity and CAPEX values, there are several other factors that contribute to the overall complexity of LNG projects. This is associated with the economy of scale, which then establishes the number of trains required to constitute the change from conventional (small) to mega or giga LNG projects. Haidar and Ellis (2010; p 1) state, "Megaprojects are unique construction projects known for their complexity, vast size, expensive cost, and long time frame compared to conventional construction projects." Love et al. (2015) propose that there is a need for future research that addresses the differences as to what complexity means for project management of projects. Furthermore, Andersen et al. (2016) suggest being cautious in cost estimates in the early phases of the project, as there needs to be an agreed threshold in each phase. If the project estimates increase twice or threefold, would the project be in a situation of Go-No-Go before FID? In addition, Andersen et al. (2016) question what a realistic initial estimate would be for the project if the concept phase was omitted and how this would affect each phase going forward. This would or should be detected with better alignment from due diligence and strong combined governance (corporate and project).

The author's observation is that from the correlation of the research, there is a trend that shows project managers are still benchmarking each phase as they would a smaller mega project, especially the Develop and Execute phases that most teams agree have unrealistic timeframes (refer to Figure 6.2).

Figure 6.2 also shows the current durations for each phase that some organisations still believe are adequate for undertaking mega (complex) LNG projects. However, this is an archaic perspective which can reflect an outdated behaviour or opinion, potentially preventing them from recognising or understanding the complexities that are associated with developing mega/giga (complex) LNG projects.

From the author's observation and suggestions acquired from research undertaken, it identifies that there was a lack of skilled management to provide direction and leadership in the Develop and Execute phases of the hypothetical case study projects as mentioned in Chapter 5.

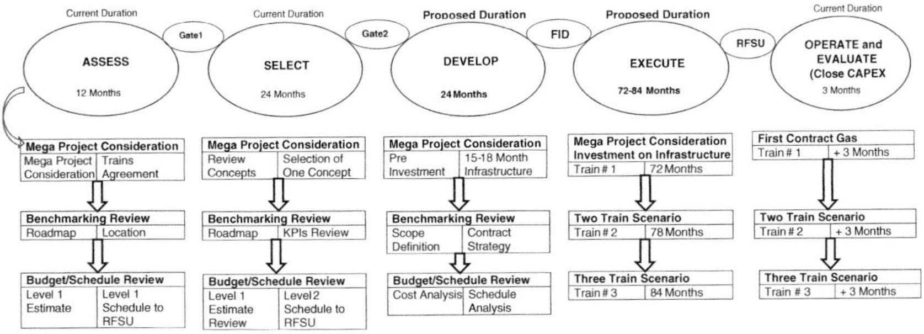

Figure 6.2 Framework for Mega LNG Projects

Accordingly, this has identified that those organisations undertaking these massive projects should rethink who has the right competencies and knowledge to manage them.

Observations/trends would suggest that the following duration changes, as illustrated in Figure 6.3, can assist in the formulation of benchmarking precedence for future mega/giga (complex) LNG projects. Haidar and Ellis (2010; p 6) state, "Throughout the construction management literature, different case studies and researchers have determined several mega project problems and causes of these problems."

Figure 6.3 outlines the two areas of change that increases the durations of the Develop and Execute phases, as also depicted in the comparisons in Table 6.1 – Life-cycle Phase Durations (mega/giga).

The information in Figure 6.3 identifies that the Develop and Execute phases for giga projects do not provide sufficient time in which to:

a) define the complex scopes, design criteria, and understand the complexities that need to be incorporated during the Develop phase;
b) freeze the design at FID and not to bring on innovative technologies after this point;
c) ensure fabrication drawings, without holds, are available to fabrication yards;
d) establish a strong and skilled project (leadership) team that is aligned with local conditions and resources;
e) instil a culture of trust between the project management teams (owner) and EPCM contractor; and
f) achieve a realistic RFSU date instead of forcing unrealistic stretch targets.

These observations are not restricted to the overall needs but are some of the significant issues that continually contribute to schedule and cost overruns in the fabrication yards and the site construction teams of the LNG process plants and infrastructures. Addressing these challenges requires a comprehensive review of project timelines and resource allocation strategies.

Table 6.1 Life-Cycle Phase Durations (Mega/Giga)

Current (mega) to Proposed (giga) phase durations		
	Current Durations	*Proposed Durations*
Assess **(initiate)**	12 months	12 months
Select **(pick)**	24 months	24 months
Develop (mature)	**18 months**	**24 months**
Execute (activate)	**50–70 months**	**72–84 months**

Authors Research

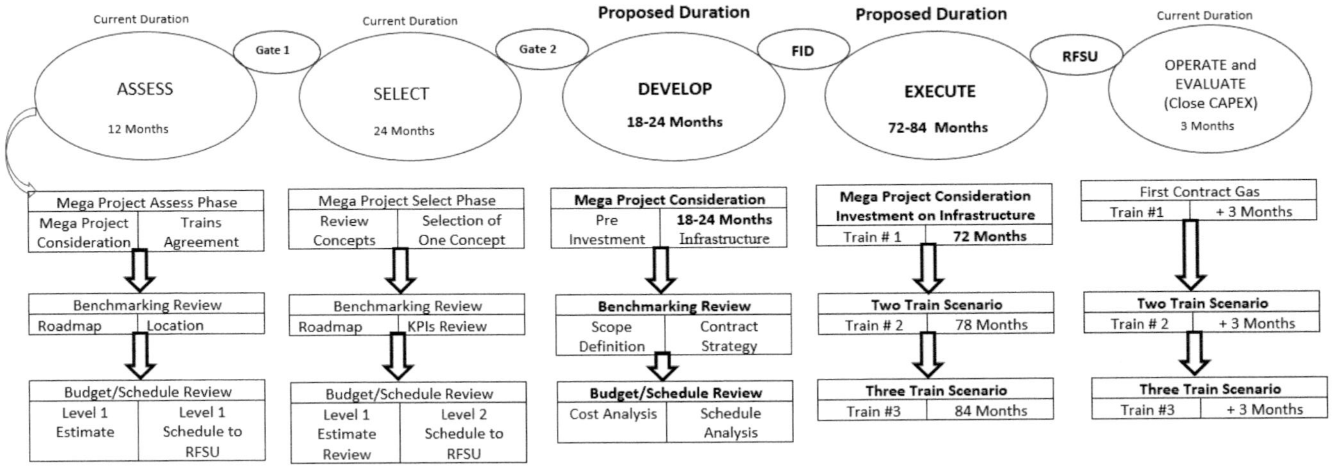

Figure 6.3 Giga Project – Recommended Benchmark Phase Durations

Authors Research

6.6 Project Life-Cycle

The majority of organisations use similar methodologies with minor name differences for each phase of their project life-cycles. This consistency facilitates communication and collaboration across the industry, allowing for shared best practices and lessons learnt.

However, it is suggested that by doing so they omit the Operate/Evaluate stage, which, for mega (complex) LNG projects, is around a 3-month duration post-RFSU till first contract gas is ready for shipping before closing the CAPEX budget. When the Execute phase is completed, so too is the actual CAPEX expenditure work and handed over to the operations team where the OPEX expenditure is then undertaken. Mukherjee and Chatterjee (2015) point out that stage-gated portioning is a key determinant to the structure across the sponsors, from a perspective of project execution, which can determine the speed and accuracy of the decision-making.

For the purpose of this book, a brief overview of the five phases of the project life-cycle is given; however, the research concentrates on the CAPEX expenditure of the four main phases.

It is crucial for mega LNG projects to implement a rigorous benchmarking analysis process within their Project Management Plan (PMP). This process should be initiated and actively supported by senior management executives throughout each phase of the project. Benchmarking allows project teams to compare their performance, processes, and outcomes against industry standards and best practices. By doing so, it enables them to identify areas for improvement, mitigate risks, and optimise project delivery. The involvement of senior management ensures that benchmarking is prioritised and that insights gained are effectively leveraged to enhance project success and efficiency. Trimble and Lester (2014) propose that it is important that the PMP is revised and, if there are any subsequent changes, it is essential that it be amended as soon as changes become apparent to ensure every member of the project team is immediately aware of the latest revision.

Conducting multifaceted workshops on benchmarking methodologies throughout the entire project life-cycle is essential. These workshops should be comprehensive, covering various aspects such as performance metrics, best practices, industry standards, and technological advancements relevant to each phase of the mega LNG project.

By engaging in these workshops, it helps to foster a deeper understanding of industry benchmarks, enabling project teams to optimise timelines, allocate resources effectively, and enhance project outcomes. This includes determining realistic durations for each phase, identifying potential bottlenecks or challenges, and establishing benchmarks for the overall project timeline and deliverables. According to Tanaka (2014), managing complex mega projects involves numerous interactions of complexity which requires project management beyond traditional project management, as the conventional project governance strategies cannot deal with multi-objectives and multilayers of stakeholders. Moreover, the workshops foster collaboration and knowledge sharing among team members, stakeholders, and industry experts. This collaborative approach ensures that the benchmarking process is thorough and aligned with the project's objectives and strategic goals.

Overall, integrating multifaceted benchmarking workshops into the project management plan enhances decision-making capabilities, improves project planning and execution, and ultimately contributes to the successful delivery of complex mega LNG projects.

Engaging JVPs or stakeholders early and effectively is crucial to ensure their understanding and support for the complexity outlined in the SOW. By increasing interaction with JVPs through clear communication, interactive workshops, emphasising benefits, seeking alignment, and securing commitment, project teams can ensure that stakeholders appreciate and support the complexity outlined in the SOW. This proactive approach lays a solid foundation for successful project execution and stakeholder satisfaction. Trimble and Lester (2014) point out that, as the word cycle implies, the phases may have to be amended in terms of content, cost, and duration as the latest information becomes apparent and is fed back to the sponsors. This provides visibility to the JVPs or stakeholders for the need to pre-invest at an earlier stage to build the infrastructure and bring on contractors to deliver these complex LNG projects. McKenna et al. (2006) point out that mega projects involve complex commercial arrangements between numerous companies and, accordingly, quick decisions from JVPs are not forthcoming, hence slow responses that can affect the schedule.

Increased market assessments necessitate a wider spread analysis of the overall market situation on projected infrastructure and qualified resource availability. This kind of assessment is crucial for understanding how market conditions are evolving and how they might impact an organisation's business or project. This would require an enhanced alignment and interaction with other mega/giga LNG project organisations throughout Australia and worldwide. According to Merrow (2011), the underperformance of mega projects requires organisations to focus on its tools and techniques to better realise the challenges and how to cope with them when developing mega projects.

6.7 Life-Cycle Phases

As previously mentioned, the development and operation of an LNG project typically go through several life-cycle phases, and as described in this book, characteristically, there are five phases. These phases outline the key stages, deliverables, and activities involved in bringing an LNG project from conception to RFSU and project handover. These five phases provide a structured approach to project management, helping organisations to plan, execute, and evaluate their initiatives effectively. Customising the framework to fit specific methodologies and organisational requirements, enhancing its practical relevance and effectiveness in supporting successful project management and strategic decision-making processes. These phases are often used to guide a project from its initial idea to its completion and ongoing evaluation. Explained below are the four main phase from Assess (initiate) to RFSU and a fifth phase of CAPEX closeout.

6.7.1 Assess Phase (Phase 1-Initiate)

The Assess phase (also known as the Initiate Phase) is typically the initial stage where the project's feasibility and viability are evaluated. This phase sets the foundation for the entire project life-cycle. The Assess (Initiate Phase) is critical as it lays the groundwork for the project's success by ensuring the project aligns with organisational goals, is feasible, and has the necessary support and resources. It sets the stage for more detailed planning and execution in subsequent phases of the project. Traditionally, at this stage, a benchmark process is carried out and a review is conducted to scope the initial options and value drivers that can be considered. This phase, often called the conceptualisation or concept development phase, is focused on identifying and reviewing alternative concepts or ideas that could potentially be developed further. This phase is crucial as it sets the stage for selecting the most promising concept to move forward into more detailed scrutiny in the subsequent Select Phase (Phase 2). By effectively conducting the conceptualisation phase, project teams can identify and refine viable concepts that have the potential to meet project objectives and deliver value. This phase prepares the groundwork for a more detailed evaluation and selection process in the subsequent phases of the project life-cycle, particularly in the Select phase where one concept will be chosen for further development and implementation. In summary, the conceptualisation phase acts as a critical bridge between initial idea generation and detailed evaluation. It ensures that only the most promising concepts are advanced into the next phases of the project life-cycle, where more rigorous analysis, planning, and implementation strategies are developed. This systematic approach increases the likelihood of successful project outcomes by focusing efforts on concepts that are both feasible and aligned with project goals. McIntosh (2018; p 11) states, "In this stage, several options for the project are considered and evaluated utilising technical and commercial criteria, and which have considered all risks that could materialise during project execution." It is in this phase where options are selected that can be benchmarked and are aligned with the project value drivers for development. In some cases, two or more options are rolled over into the Select phase at which time a design completion may be used to decide which option should be used going forward. At the end of this phase, there should be a clear understanding of why the project is necessary and what it aims to achieve.

For these proposed mega/giga LNG projects, it is recommended at this phase that the level 1, 2, and 3 estimates should be developed for the entire life-cycle. To do so requires an in-depth understanding of the overall Scope of Works (SoW) in this phase. Traditionally, this is not available so early; however, with the early engagement of construction and IM in this phase, it can be made available even if the Pareto rule of 80/20 is applied. Raydugin (2013) proposes that any project objectives or baselines, such as base estimates or schedules, are models that mimic the future project reality. Accordingly, the development of a robust project assurance plan needs to be completed and approved. This forms part of the Decision

Support Package (DSP) which is developed as early as possible within the Assess phase and refined during the subsequent phases. The DSP is a crucial component developed early in the project life-cycle, typically within the Assess Phase, to support informed decision-making throughout the project. By developing the DSP early in the Assess Phase and refining it throughout subsequent phases, project teams can effectively manage complexity, mitigate risks, and enhance the likelihood of project success. The DSP acts as a critical tool for aligning stakeholders, guiding decisions, and maintaining clarity and transparency throughout the project life-cycle.

The overall conceptual roadmap outlines the strategic framework for the project, detailing key milestones, deliverables, and the ownership assigned to each phase throughout its life-cycle. This structured approach ensures clarity on responsibilities and timelines, facilitating efficient progress tracking and alignment with project goals. This 12-month phase is dedicated to completing the framing process and preparing a robust business case. The relatively short timeframe underscores the urgency and focus required to achieve these objectives swiftly. By the end of this phase, the aim is to finalise the project's scope, objectives, and strategies, ensuring they align closely with organisational goals. Simultaneously, a comprehensive business case will be developed, which will undergo rigorous evaluation for endorsement or, if necessary, refinement and resubmission. This approach allows for the maintenance of agility while ensuring that decisions are well informed and aligned with stakeholders' expectations. In the event that the business case is rejected or requires recycling, the advantage of this 12-month timeframe is that it ensures only a minimal amount of funding is consumed. This approach safeguards financial resources while allowing for timely adjustments and improvements based on feedback and evaluation. It emphasises prudent resource management and the ability to pivot effectively, ensuring that the project remains aligned with strategic objectives without compromising financial stability. It is recommended that the Key Deliverables for the Assess Phase be in the following main categories to ensure comprehensive evaluation and planning (reference to Table 6.2 for details):

- General – depicts the development of the main processes, procedures, and the DSP, and provides a comprehensive overview of the development of these processes and the DSP process.
- Joint Venture Partners (JVPs) – depicts what is required by the JVPs. When outlining what is required by JVPs, it is important to detail both the general expectations and specific responsibilities that each party must fulfil.
- Technical – depicts the technical requirements for the phase. Defining the technical requirements for a specific phase is crucial to ensuring successful execution and meeting project goals. By outlining these technical requirements clearly, ensures that all stakeholders have a comprehensive understanding of what is needed for the successful execution of the technical phase.

Table 6.2 Assess Phase Key Deliverables

Update Master Schedule (Upstream/Downstream)
GENERAL
Develop Project Life-cycle Road Map
Develop Project Charter
Develop Primary Project Management Plan (PMP)
Develop Preliminary Project Execution Plan (PEP)
Develop Preliminary Risk Register
Develop Preliminary Action Tracking System
Develop Preliminary Interface Management (IM) Plan
Develop Preliminary Communication Plan (CP)
Preliminary Review of Infrastructure Requirements
Review Lessons Learnt in this Phase
Joint Venture Partners (JVPs)
Preliminary Alignment of JVPs
Preliminary Strategy for Government Support and Approvals
Infrastructure Support Review
Workshop for preliminary project Management Plan (PMP)
Discussion on Pre-investment into early Infrastructure
Commercial/Marketing
Ascertain Market Viability
High-Level Infrastructure/Resource Availability Review
Technical
Preliminary Land Acquisition Review (if Required)
Develop Preliminary Short List of Concepts
Develop Level 1 Budgets and Schedules
Develop Level 1 Schedule Risk Analysis
CHECK POINT 1 REVIEW (Schedule/Cost/Quality)
Benchmarking
Develop Preliminary Benchmarking Analysis
Update Master Schedule (Upstream/Downstream)
Preliminary Benchmarking Workshop (includes JVPs)

Note: following the Assess phase, most large oil and gas (LNG) projects are divided into two groups that of the Upstream and Downstream (offshore/onshore) teams.

Table 6.3 Select Phase Key Deliverables

SPLIT UPSTREAM/DOWNSTREAM INTO TWO SEPARATE TEAMS
GENERAL
Develop Sand-alone Reference Case
Develop Contracting Strategy
Review Preliminary Project Execution Plan (PEP)
Review Life-cycle Roadmap for Scope Changes
Update the Project Management Plan (PMP)
Update the Risk Register
Update the Interface Management Plan (IMP)
Update the Communications Plan (CP)
Revise Infrastructure Plan Requirements (if any)
Review Lessons Learnt in this Phase

(*Continued*)

Table 6.3 (Continued)

Joint Venture Partners (JVPs)
Review meeting with Stakeholders to agree requirements
Review Draft Strategy for Government Approvals
Joint Venture (JVPs) Alignment Session
Update on Infrastructure Support Review Workshop
Review Workshop for Project Management Plan (PMP) approval
Workshop to Review Budgets and Schedule Proposals
Review the Develop Phase Schedule and Cost Proposals
Review of Risk Analysis for the Develop Phase
Commercial/Marketing
Develop Market Strategy
Workshop on Technical and Commercial Risks
Workshop on Availability of Skilled Workers (LNG)
Technical
Develop Land Acquisition Concept
Develop Business Case Proposals
Develop Range of Concept Selection (Screening)
Update Workshop on Schedule Risk Analysis
Field Appraisal Review (Upstream Specific)
CHECK POINT 2 REVIEW (Schedule/Cost/Quality)
Benchmarking
Review and Update Benchmarking Durations
Conduct Benchmarking Workshops Split Upstream and Downstream

Authors' Research

- Commercial/Market – depicts the commercial and market requirements. Defining commercial and market requirements is essential for ensuring that a project aligns with market needs and business goals.
- Benchmarking – depicts the benchmarking processes and requirements for this phase and the next phases to reach a predicted RFSU date. Benchmarking processes and requirements are essential for assessing performance, setting standards, and guiding improvements throughout the project life-cycle. This ensures that the project is evaluated comprehensively and systematically, paving the way for a successful transition to RFSU and aligning with overall project goals.

These high-level requirements should be the minimum prerequisite to manage the Assess phase of a mega (complex) LNG project, providing a solid foundation for decision-making and subsequent phases of the project life-cycle.

For this phase of mega/giga LNG projects, it is recommended that a more in-depth process be developed in this early stage, which is associated with managing these complex projects at the beginning of their project life-cycle. By implementing a thorough and structured approach in these early stages, mega or giga LNG projects can mitigate risks, optimise resource allocation, and enhance overall project outcomes. This disciplined approach not only supports successful project execution but also fosters long-term operational efficiency and sustainability.

As a result, building upon the need for a robust methodology is provided for this phase, including how the team needs to be managed as the project progresses into subsequent phases. In the context of mega or giga LNG projects, it is crucial to empower the team to emphasise and expedite the early development of processes, methodologies, and team structures. Empowering the team to prioritise and accelerate the early development of processes, methodologies, and team dynamics is a strategic advantage in mega or giga LNG projects. Traditionally, these elements might only receive attention later in smaller projects, but their proactive integration from the outset enhances project efficiency and resilience. This, as with everything else concerning the management of mega/giga (complex) LNG projects, requires organisations/management to undergo a fundamental mindset change at this early stage regarding how this and all other phases should be developed and managed. This approach ensures that the project team is well prepared to navigate complexities and capitalise on opportunities from the outset of mega or giga LNG projects. Haidar and Ellis (2010; p 6) state, ". . . underestimation of size and complexity, non-realistic planning, inadequate project organisation, inefficient structure, and poor integration – highlight the fact that conventional management practices are insufficient to handle mega-projects characteristics." Consequently, when embarking on mega or giga LNG projects, it becomes imperative for organisations to adapt their approach to developing team structures and engaging experienced individuals across different phases of the project. When managing mega or giga LNG projects, project managers often need to adjust from the traditional organisational structures and focus composition and, accordingly, require to be more focussed on the overall outcome. The shift towards a more outcome-focused approach in managing mega or giga LNG projects necessitates higher competency testing of personnel. In response to the heightened focus on achieving successful project outcomes in mega or giga LNG projects, there is a critical need for rigorous competency testing of personnel. This ensures that individuals involved in mega LNG projects possess the necessary calibre to execute their roles effectively across all phases. This approach verifies that team members have the requisite skills, knowledge, and expertise to successfully conduct each stage of the project. Simmons et al. (2020; p 4) state, ". . . it is important to measure perceptions of the industry as it presently exists to identify benchmark conceptualisations of professional competencies within construction." By rigorously testing competency, organisations can mitigate risks associated with inadequate skills and knowledge gaps among project personnel. This proactive approach not only enhances project performance and efficiency but also contributes to the overall success and sustainability of mega or giga LNG projects.

By having an improved selection of personnel and processes at this early stage and providing a better understanding of what a giga project is, enhances the appreciation of the magnitude of the giga LNG projects. This methodology underscores the importance of aligning personnel capabilities with project requirements, ensuring that the team is well equipped to achieve successful outcomes across all phases of complex LNG projects.

Moving forward, mega LNG projects necessitate organisations to exercise greater diligence in selecting and appointing project managers and teams. This ensures they are well equipped to handle the complexities inherent in such large-scale endeavours. Consequently, achieving success in completing multifaceted LNG projects requires organisations to adapt their management styles, concepts, and ethos. This adaptation is essential to ensure projects are delivered on time and within budget, considering the intricate nature of these endeavours. Raydugin (2013) suggests that it is not possible to manage factors influencing project outcomes without understanding their definition. Such a mindset change requires the organisation to stop, reconsider, and reevaluate their management styles to be able to undertake such a momentous departure from the conventional project management process. This requires strong management control and commitment to the re-evaluation whereby this new concept can be achieved. Enhancing the management concept for complex LNG projects undoubtedly brings challenges, prompting a critical examination of the need to move away from or modify the status quo. By questioning the status quo and embracing progressive management concepts tailored to the complexities of LNG projects, organisations can enhance project resilience, efficiency, and overall success. This transformative approach not only addresses current challenges but also positions projects for long-term sustainability and competitiveness in the global energy landscape. This exploration highlights the importance of challenging conventional practices to optimise management strategies for complex LNG projects. Tanaka (2014) suggests that project management needs to be understood as a complex discipline as it involves complexity, uncertainty, and ambiguous reality.

Many organisations and project managers may hold the belief that their current management processes are adequate for delivering mega or giga projects. To change, or consider a change, from the conventional project management structures necessitates the commitment of senior management/stakeholders into the early development of new and reformed processes, which then become live documents throughout the project life-cycle. In essence, engaging senior management and stakeholders early in the development and reform of project management processes establishes a solid foundation for successful implementation and sustained improvement over time. It is a strategic approach that recognises the importance of leadership in driving organisational change and achieving project success. Tanaka (2014; p 66) states, ". . . management of these projects should be referred to as 'program management' as managing complex projects should be distinguished from project management focusing on delivery which is mostly based on positivist management processes and operational techniques." The concept for managing mega/giga (complex) LNG projects is to establish and agree upon robust and live PMP and PEP documents that become the key, early on, to the success of these projects. They provide a structured framework for managing complexities, mitigating risks, optimising resources, and ensuring effective stakeholder engagement and communication. This disciplined approach enhances the likelihood of delivering the project on time, within budget, and to the desired quality standards.

Trimble and Lester (2014) point out that the PMP lists the phases and encapsulates all the main parameters, standards, and requirements of the project in terms of time, cost, and quality/performance by setting out the 'Why', 'What', 'When', 'Who', 'Where', and 'How' of the project. Building on the foundation of robust PMP and PEP for managing mega/giga LNG projects, further prominence should be placed on several key aspects to enhance project success. This concept requires the establishment of a robust Communications Plan (CP), Decision Support Packages (DSPs), and the introduction of a robust Interface Management (IM) plan. Integrating these into the project management framework enhances the overall effectiveness and resilience of mega/giga LNG projects, contributing to their successful delivery and long-term sustainability. Managing giga/mega LNG projects involves intricate methodologies and interface arrangements to ensure successful execution. Given the complexity, it is crucial that Project Management Professionals (PMPs) overseeing these projects implement a robust benchmarking analysis process. This process should be initiated and strongly supported by senior management executives throughout each project phase. By incorporating these elements, PMPs can navigate the complexities of mega/giga LNG projects more effectively. The benchmarking analysis process, underpinned by senior management support, serves as a crucial tool for achieving project success and delivering value to stakeholders.

During the Assess phase, especially in the context of mega/giga LNG projects, is where the development of a preliminary benchmarking analysis is crucial. By undertaking a comprehensive benchmarking analysis during the Assess phase, PMPs lay the foundation for informed project management decisions. This proactive approach helps in managing risks, optimising performance, and ultimately achieving successful outcomes in mega/giga LNG project developments. Refer to Table 6.2 for the key deliverables for the Assess phase, grouped into the following areas – General, Joint Venture Partners, Technical, Commercial/Marketing, and Benchmarking.

The following Tables 6.2, 6.3, 6.4, and 6.5 are the recommended main Key Deliverables for each of the four phases for the mega/giga LNG project. Also refer to attachment A – Proposed Roadmap.

6.7.2 Select Phase (Phase 2-Pick)

Traditionally on mega/giga (complex) LNG projects, the Select phase (phase 2) consists of Upstream and Downstream teams with an overarching Project Director to interface between these teams. McIntosh (2018; p 16) states, "Typically in the Select Stage, there will be a deep dive into all the elements described in the Appraise Stage, but in addition, the Project business objectives, vision and mission, and the final selection of Project Drivers are detailed." According to Tanaka (2014), the Select phase is when there is an in-depth examination conducted on the options or concepts that were carried through from the Assess phase. The objective of this phase is to screen each of the options to identify the optimum development concept which can ascertain a standalone reference case for the proposed Upstream and

Downstream developments. Depending on the size and complexity of the project, the downstream portion is split into two further segments, i.e.:

- Group A – Analysis; and
- Group B – Business case selection.

The reasoning for this is due to the size and complexity of these mega/giga LNG projects, and especially for the downstream portion, it may not be clear which option should be carried into the Develop phase. Therefore, the Select phase for these types of mega/giga projects may take longer to reconcile on a preferred reference case to bring into the Develop phase. It is essential that, at this point, benchmarking data is reviewed.

The objective of this book is to understand the downstream portion of these mega/giga projects to increase the in-depth knowledge base that is necessary to develop a robust reference case. The Select phase needs to have a substantial update of all processes and procedures, such as the PMP, PEP, Risk Register, IM plan, Communications Plan, and the Baseline Schedule. Along with this, more emphasis must be placed on long-lead items, not only procurement but also the infrastructure requirements.

Consideration should be given to the fact that, within the LNG industry, the current standard contract-type strategies used on smaller oil and gas projects seem to be too loose or inadequate to manage mega/giga (complex) LNG projects. This is evident when as these contract values increase with variations due to lack of scope definition, schedule slippages and costs escalation occur. With giga LNG projects, there is a need for a change in thinking in the way the contract strategies are developed. This change requires a novel approach in determining the appropriate strategy for the downstream portion, with more cross-sectional involvement and engagement of the JVPs and possible major contractors (incentive contracting) who are capable of undertaking an Engineering, Procurement, Construction, Management (EPCM) contract. Merrow (2011; p 133) states, "Incentive contracting involves sharing cost overruns between owners and contractors." It is recommended that, in order to achieve this mindset adjustment, a series of in-depth workshops at the beginning of the Select phase should be undertaken to consider the best contracting strategy to be used. It would also be recommended that, at this stage, there is a need to engage EPCM contractors. Refer to Table 6.3 for the key deliverables for the Select phase, grouped into the following areas – General, Joint Venture Partners, Technical, Commercial/Marketing, and Benchmarking.

6.7.3 Develop Phase (Phase 3-Mature)

The Develop phase consists of two principal areas for each of the Upstream (offshore) and Downstream (onshore) portions as follows:

- Basis of Design (BoD);

- The Basis of Design (BoD) is a critical document in the design and engineering processes, particularly in fields such as architecture, construction, and systems engineering. It provides a clear framework and rationale for the design decisions and solutions proposed for a project.

- Front-End Engineering Design (FEED).

    - Front-End Engineering Design (FEED) is a crucial phase in the development of a project, particularly in industries such as oil and gas, petrochemical, and large-scale infrastructure. It serves as a bridge between the conceptual design phase and the detailed engineering phase.

In this phase, in-house consultants/contractors who assisted in an integrated manner to finalise concept selection and the development work processes support the owner's project team. During the Develop phase, an EPCM contract is tendered, ready for an award at FEED, for an Engineering Management Services (EMS) contractor to form part of an Integrated Project Management Team (IPMT) for the FEED portion of the Develop phase and for the Execute phase. McIntosh (2018; p 18) states, "The main focus of the Define stage is to complete the various areas of work performed in the Select stage and finalise all documentation required to support project sanction."

During this phase, the project schedule is further developed to update the WBS to include the list of long-lead item requirements to allow for pre-investment prior to FID. Refer to Table 6.4 for the key deliverables for the Develop phase, grouped into the following areas – General, Joint Venture Partners, Commercial/Marketing, Technical, and Benchmarking.

Table 6.4 Develop Phase Key Deliverables

PHASE 3 – DEVELOP PHASE
Key Deliverables
DOWNSTREAM
GENERAL
Workshop for Updating Contract Strategy
Review and Update the Project Execution Plan (PEP)
Review and Update the Project Management Plan (PMP)
Review and Update the Risk Register
Review and Update the Information Management Plan (IMP)
Review and Update the Communications Plan (CM)
Review the need for Purchasing Long Lead Items
Development of Change Management Process for the Execute Phase
Joint Venture Partners (JVPs
JVP Alignment Workshop
Workshop to Align Government Support and Approvals Process
JVP Alignment Session on the Basis of Design (BoD)
JVP Alignment on the Front End Engineering Design (FEED) Process
JVP Alignment on the PMP Update
JVP Review of Cost and Schedule for the FEED
FINAL INVESTMENT DECISION (FID)

(*Continued*)

Table 6.4 (Continued)

Commercial/Marketing
Workshop on Technical and Commercial Risks for the Execute Phase
Workshop on Availability of Skilled Workers (LNG) for the Execute Phase
Technical
Basis Of Design (BoD) Preparation
BOD Workshop Alignment
Workshop on Schedule Risk Analysis for the Execute Phase
Field Appraisal Review (Upstream Specific)
CHECK POINT 3 REVIEW (Schedule/Cost/Quality)
FEED Study Workshop (Study Analysis and Process)
CHECK POINT 4 REVIEW (Schedule/Cost/Quality)
Benchmarking
Benchmarking data review for the Execute Phase

Authors' Research

6.7.4 Execute Phase (Phase 4-Activate)

The Execute phase is where the work program and Authorities For Expenditure (AFE) are to be submitted to the JVPs for approval once the project has received FID approval. It should be noted that between FID and JVPs approval there would potentially be a need to obtain bridging funds to commence the design activities whilst awaiting JVP approvals. McIntosh (2018; p 21) states, "The activities in this stage typically . . . include those that are either the responsibility of the EPC contractor or the Owner's project leadership team."

In the first 2–3 months of the execute phase, the awarding of all contracts is finalised. The level 4 and level 5 schedules are also finalised, and all reporting requirements agreed and approved by the owner's team. The completion and freezing of the design drawings are accomplished, and the final constructability review workshop is undertaken, and all regulatory approvals received. Refer to Table 6.5 for the key deliverables for the Execute phase, grouped into the following areas – General, Joint Venture Partners, Commercial/Marketing, Technical, and Benchmarking.

6.8 Operate/Evaluate Phase (Appraise/CAPEX Close-Out)

The Operate/Evaluate phase in project management encompasses two main aspects: operating the project's outputs and evaluating the project's overall performance. The key deliverables and activities associated with this phase are as follows:

6.8.1 Operate Phase Key Deliverables

- **Operational Plan**: This document outlines how the project's outputs will be integrated into ongoing operations. It includes details about roles and responsibilities, maintenance procedures, support mechanisms, and any necessary training for personnel.
- **User Documentation**: User manuals, training materials, or other documentation that helps end-users understand how to use and maintain the project's deliverables.

Table 6.5 Execute Phase Main Key Deliverables

DOWNSTREAM
GENERAL
Budget Reviews and Final Analysis
Update Actual versus Planned
Weekly Schedule Forecasts
Weekly Cost (Budget) Forecast
Monitoring of Change Management Register
Monitoring of Risk Management Register
Final Review and Update of the Roadmap
Lessons Learnt
Joint Venture Partners (JVPs)
JVP Alignment Workshop for Risk Management
JVP final Review of Schedule and Budget
Asset Acceptance End of CHEK POINT 5
Commercial/Marketing
Continuous Review of the Market Strategy
Technical
Design Freeze at the Start of the Execute Phase
Alignment of Upstream and Downstream Teams
CHECK POINT 5 REVIEW (Schedule/Cost/Quality
Benchmarking
Benchmarking Review Workshop (Includes JVPs) Discussions on what went well and what went wrong (if anything)

Authors' Research

- **Transition Report**: A report that summarises the transition process from project implementation to operations. It may include an assessment of any challenges faced during the transition and how they were addressed.
- **Support and Maintenance Procedures**: Clear procedures and guidelines for ongoing support and maintenance of the project's deliverables. This includes schedules for regular maintenance, troubleshooting processes, and escalation procedures for issues.
- **Monitoring and Feedback Mechanisms**: Establishing systems for monitoring the performance and usability of the project's outputs in the operational environment. This includes feedback loops, surveys, or other methods to gather input from end-users.
- **Quality Assurance and Quality Control Processes**: Procedures for ensuring that the project's outputs continue to meet quality standards and performance expectations in the operational phase.

6.8.2 Evaluate Phase Key Deliverables

- **Project Evaluation Report**: A comprehensive report that assesses the project's overall success, including whether it met its objectives, stayed within budget, and adhered to the timeline.

Table 6.6 Operate/Evaluate Phase Main Key Deliverables

READY FOR START-UP (RFSU)
GENERAL
Lessons Learnt
Final Budget Review
Final Schedule Review
Closeout of Capital Expenditure (CAPEX) Budget
Joint Venture Partners (JVPs)
JVP Review of Budget and Schedule Status
Commercial/Marketing
Technical
Punchlist Items for any Carryover Work (CoW)
Closeout Reports to Include Upstream and Downstream Portions
Benchmarking
Benchmarking Final Alignment Review

Authors' Research

- **Lessons Learned Documentation**: A document that captures insights and lessons learned throughout the project's life-cycle. It highlights what went well, what could have been improved, and recommendations for future projects.
- **Post-Implementation Review (PIR)**: An evaluation of the project's performance after it has been operational for a defined period. It compares the expected outcomes with actual results and identifies any discrepancies.
- **Performance Metrics and KPIs**: Analysis of Key Performance Indicators (KPIs) and metrics established during the project's planning phase to measure its success. This can include financial metrics, customer satisfaction scores, or any relevant performance indicators.
- **Stakeholder Feedback**: Gathering feedback from project stakeholders, including end-users, team members, and sponsors, to gauge their satisfaction and obtain insights into areas for improvement.
- **Recommendations for Continuation or Closure**: Based on the evaluation findings, recommendations are made regarding whether the project should continue, be expanded, or be terminated. This may also include suggestions for enhancements or follow-up projects.

The Operate/Evaluate phase is essential for ensuring that the project's outputs continue to add value and for capturing lessons learnt that can inform future initiatives. The key deliverables in this phase contribute to ongoing success and the continuous improvement of project management processes within an organisation.

6.9 General Overview

It is necessary that the due diligence process at the end of each phase or checkpoint is undertaken, as it is crucial for ensuring that the project team is adequately prepared to progress to the next stage. This independent evaluation helps verify that all required deliverables have been completed satisfactorily, potential risks have been identified and mitigated, and that the project aligns with its overall goals and objectives. It provides a structured opportunity to review and validate the project's progress before committing to the next phase, thereby enhancing the project's chances of success.

Overall, due diligence is an essential part of project management that helps safeguard against potential risks and uncertainties, ensuring that projects remain on track and ultimately achieve their objectives. Kerzner (2014) proposes that project success can be measured intermittently throughout the phase or gate review processes that are part of the project management methodology. This assurance process or due diligence process is conducted on cost/schedule robustness, understanding, and preparation of the SoW, and how these types of issues are managed. From this assurance check, the decision is made, sanctioned or not, by the stakeholders to continue into the next phase or recycle or stop the project. According to Kerzner (2014), certain questions that should be asked are:

- was management kept informed as the project progressed;
- were the proper resources assigned and were they efficiently utilised; and
- is the correct methodology being used?
- The essential points to take note of are:
- during the Assess, Select, and Develop phases, has the leadership team chosen the correct project(s) to go forward with; and
- in the Develop phase, is there enough detailed design and scope information to allow the leadership team to be confident to undertake the FID approval process.

Figure 6.4 shows the process integration of the project life-cycle and how these phases and checkpoints form part of a project benchmarking process.

McIntosh (2018; p 10) states, ". . . the role of the gatekeeper, which is more critical in the early stages of the project up to project sanction, thereafter, reducing but still present to the end."

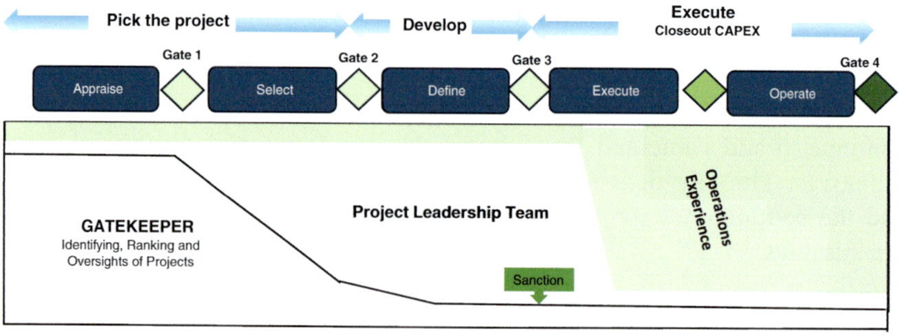

Figure 6.4 Integrated Involvement throughout the Project Life-Cycle

(McIntosh 2018; p 10)

The governance process is a function that provides the stakeholders with confidence that the project team is on track and, consequently, the project is still viable to continue. According to Cooper (2008), stage gates or checkpoints form a tough post-launch review to foster accountability for results of the phase undertaken and governance for continuous improvement.

There are normally three main stage gates or checkpoints prior to the approval to enter the execute phase where the main decision (go/no go/hold/recycle) is made to approve (or not) the Financial Investment Decision (FID). On the approval of FID, the project is at a point of no return. These gates or checkpoints, as shown in Figure 6.4, are specifically at the end of what Merrow (2011) calls the Front-End Loading (FEL) phases 1, 2 and 3. Therefore, when FID is approved, which is at the end of FEL3, the project is at a point of no return as it moves into the Execute phase. Winch (2010) suggests that the project life-cycle depends on information, feedback, and control, without this, it can become a problem when determining the project life-cycle. If information loops fail, the project tends to go over budget and schedule.

As previously stated, during the Assess phase the project roadmap is developed which depicts the life-cycle durations of each specific phase of the project, including operate/evaluate. This provides a framework, at an extremely prominent level, for the deliverables as dot points, and responsible managers assigned to the phases. Durations are benchmarked against similar LNG projects. However, it is proposed to revisit the durations for developing mega/giga LNG projects, due to their complexity.

The common factor in all these processes is that of planning. If the planning (benchmarking) is not correct or insufficient effort has been made when baselined into the planning process, there is a knock-on effect, and the process fails. In these cases, planning sets the policies, strategies, and critical success factors with guidance from the JVPs and the project manager. Williams (2002) points out that one of the most important aspects of a project life-cycle is the project commencement. This considers the initial planning processes established by the team and common processes to follow. The project commencement is where the model is developed for the initial roadmap of the project, establishing the project life-cycle durations. For this high-level roadmap, the phase durations are benchmarked against similar LNG projects, where applicable (sound benchmarking). Williams (2002) establishes that models or roadmaps need to be revisited during each phase of the project to allow the project team to conduct rational decision-making. This would involve revisiting the benchmarking data that has been baselined in the schedule and ascertaining how the process may need to be revised, which may affect the potential moving of the end date of the project life-cycle.

The managerial difficulties increase as early as the Assess phases of the project life-cycle due to the complexity of these massive LNG projects. It is at this stage where the project team may discover or recognise the existence of Black Swans which may affect the project integrity to continue forward. Unfortunately, as observations have shown, if these potential issues (Black Swans) are not highlighted at the first risk gate, these critical issues may be lost as the project proceeds into the

Select phase, as the Assess project manager and his team normally do not continue into the following phase (Select). According to Bogle (2007), Black Swans are rare events with an extreme impact that could have been predicted. Consequently, these critical findings may not be apparent to the new team as the project moves forward with a new project manager and team. Accordingly, Cooper (2008) suggests that, after the Assess (concept) phase, projects rarely get deferred, as they are then part of the overall company road map and the overall company financial plan. Cooper (2008) goes on to suggest that all other gates after gate one (Assess to Select) are just rubber stamps. Accordingly, the first gate must not be the only fully scrutinised gate within the three-stage assessment prior to the Execute phase. As such, better due diligence should be undertaken by the corporation and project management team to ensure all gates or checkpoints receive strict compliance before moving forward. Trimble and Lester (2014) point out that the project life-cycles are control systems and provide decision stage gates at the end of each phase, at which time an independent team reviews the status of the project. This is the crossing point of the phases of the life-cycle which form opportune milestones for senior management to decide whether to proceed and provide further funding into the next phase or abort (go/no go/hold or recycle).

The author's observation suggests the context related to large-scale LNG projects, practically emphasising the need for adequate time in the Develop phase of these multidimensional mega/giga LNG projects before moving into the FID. The necessity for more time in the Develop phase for mega/giga LNG projects reflects the need to thoroughly understand and address the complexities, risks, and requirements associated with such large-scale endeavours before making critical investment decisions. By doing so this would result in the reduction or even elimination of the cost and schedule overruns presently experienced on mega (complex) LNG projects worldwide.

Reviewing lessons learnt from previous projects is a crucial aspect for continuous improvement in project management. To effectively leverage lessons learnt, project teams should establish systematic processes for documenting, sharing, and applying insights from previous projects. This ensures that valuable knowledge is captured, communicated, and utilised to optimise project performance and achieve superior results.

It would be prudent for project managers, who bring tacit information from previous projects into their current project development, to not assume that this information is not biased to the current best-in-class projects and ignore the use of quantifiable benchmarking data. By striking a balance between leveraging tacit knowledge and utilising quantifiable benchmarking data, project managers can effectively navigate complexities, mitigate risks, and optimise project performance towards achieving best-in-class outcomes. Koskinen et al. (2003; p 281) state, "Tacit knowledge represents knowledge based on the experience of individuals. It expresses itself in human actions in the form of evaluations, attitudes, points of view, commitments, motivation, etc." Specifically, that by accepting this tacit information, merely because this is better than the current benchmarking data, erodes the value of the quantifiable benchmarking data being used. When tacit knowledge from previous projects is given undue weight solely based on perceived superiority to quantifiable benchmarking data. By recognising the potential pitfalls of over-relying on tacit knowledge at the expense of quantifiable benchmarking data, project managers can strike a balance that leverages the strengths of both approaches. This approach promotes informed decision-making, minimises biases, and maximises the likelihood of achieving best-in-class project outcomes. Koskinen et al. (2003; p 285) state, "Only seldom does a single person know enough to solve complex problems faced in projects. In many projects knowing how to find and apply relevant knowledge efficiently is more practical than trying to master a large amount of knowledge."

The author's investigation has disclosed that there is mixed opinion from organisations regarding whether traditional project management processes, including benchmarking, are adequate enough for developing mega/giga (complex) LNG projects. Kozak (2004; p 8) states, "Benchmarking is an action discovering the specific practices responsible for high performance, understanding how these practices work and are achieved, and adapting them to one's organisation, while benchmarks represent performance measures such as how many, how fast, and so on."

From the author's experience and observations, the successful execution of mega and giga LNG projects requires a multifaceted approach involving meticulous planning, experienced management, effective risk mitigation, and strong stakeholder engagement. Leveraging technological advancements and maintaining a focus on financial and operational control are also crucial. By addressing these factors, stakeholders can better manage the complexities of large-scale LNG projects and minimise the risks associated with schedule overruns and other challenges. Accordingly, it is observed that there is a substantial difference between developing mega/giga LNG projects (two or more process trains) and the complexity they contain compared to traditional LNG projects (one process train). Cantoni and Pagnone (2020) imply that one of the key features of mega projects is the difference from traditional projects due to their size and complexity. Therefore, it is important to recognise that, during the phases of these massive multifaceted projects, it is not practical to manage them using traditional project management processes, as these mega projects continue to overrun on budget and time (schedule) when doing so. Cantoni and Pagnone (2020; p 1) state, "Megaprojects do have a history of failures, in terms of missed achievements of the planned objectives." The main reason traditional management processes are not adequate for mega/giga LNG projects is due to the overall complexities (multiple mega projects that exist within a giga or mega project itself) that are involved in establishing them. These complexities include (but are not limited to) such items as multiple stakeholders, environmental issues, lack of trained personnel, colossal infrastructures required (such as MOFs, causeways, extensive accommodation), complex supply chains, and supporting utilities plus material deliveries, etc. Cantoni and Pagnone (2020; p 8) state, ". . . it seems the most problematic aspect that challenges mega project management is the involvement and alignment of many stakeholders, as it was initially assessed by identifying mega projects complexity." This also includes the effective use of CIs and how they can provide an integrated approach for efficiencies in each phase of a project's life-cycle. Active CIs assist in the management of the multiple stakeholders and their expectations for sustainable outcomes. Cantoni and Pagnone (2020; p 10) state, "In terms of economic benefits that sustainable mega projects should bring, it is necessary to consider fair financial rewards and earnings for stakeholders that bring resources."

To develop a benchmarking process for mega/giga (complex) LNG projects, there is a need to understand the difference between traditional LNG projects (one process train) and mega/giga (complex) LNG projects (2, 3, or 4 process trains).

There are two types of LNG single trains. The first is normally when a site has been established with the first train and waiting for funding from the JVPs and approvals to install a second train (fabrication of modules and installation of the second train). At this stage, the infrastructure is established. Second is a single train in a small plant where the infrastructure is minimal, or already exists, i.e., roads, etc., better known as a traditional LNG project (small project), not a mega/giga (complex) LNG project. Kerzner (2006; p 329) states, "Mega projects may have a different set of rules and guidelines from those of smaller projects."

For clarity, it must be noted that benchmarking the CIs for process trains is not an issue as they are a repetitive item, like compressors and turbines, i.e., modular built and delivered to site. Arndt and Edrington (2011; p 1) state, "Benchmarking is a means of comparing performance to an established performance baseline." The things that are an issue to benchmark on mega/giga (complex) LNG projects are the bulk earthworks, foundations, infrastructure like Materials Offload Facilities (MOF), causeways, jetty, planning (vessels/barges and other supply transport), and large accommodation camps. These are determined by the location and environmental issues to support the development of these projects, notwithstanding the complex stakeholder interface/involvement. Due to these multifaceted issues, including the size/complexity of mega/giga LNG projects, they tend to consequently exceed the budget and or schedule. Ma and Fu (2020; p 2430) state, ". . . current research on the influence of complexity on project success is limited. Neither theoretical knowledge nor practical experience is sufficient to guide the management of construction projects."

Overruns in projects have been continuing for many years; however, with the introduction of mega/giga projects, these overruns are becoming more prominent compared with those identified in smaller projects. Flyvbjerg (2014; p 9) states, "Performance data for megaprojects speak their own language. Nine out of ten such projects have cost overruns; overruns of up to 50% in real terms are common . . ."

According to McKenna et al. (2006), organisations need to adopt a more global approach to project management frameworks that increases their performance and, consequently, better utilise the resource pools available, particularly with mega projects.

There is an enormous difference when it comes to budget and schedule overruns between small projects and mega projects. The scale of a project can greatly impact both budget and schedule management. For small projects, overruns are often more

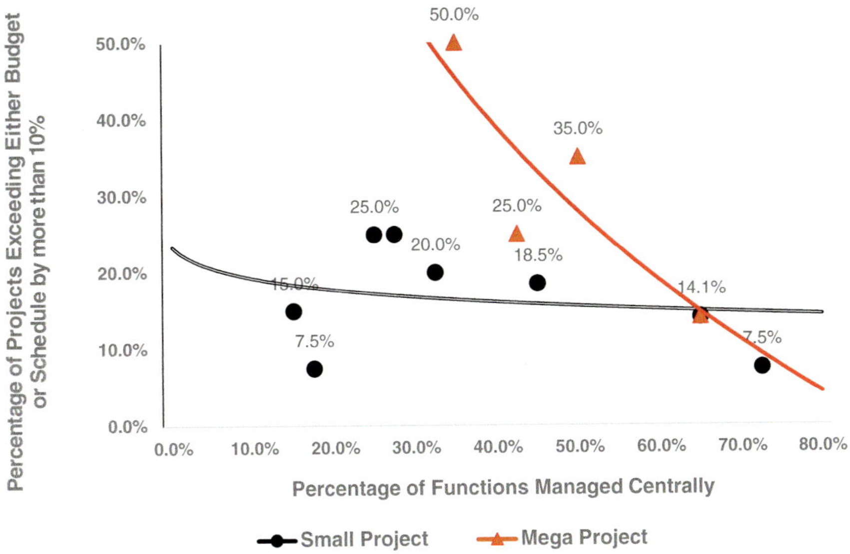

Figure 6.5 Identifies % of Project Overruns – Small and Mega Projects

Adapted by Author from McKenna et al. (2006; p 5)

manageable and can be addressed with minor adjustments. However, for mega projects, the stakes are much higher. Figure 6.5 provides the comparison of the extent of overruns that are apparent between small projects and mega projects.

All the intertwined aspects of trying to provide performance measures, such as CIs, need to be considered to develop a solid benchmarking concept and a benchmarking strategy for mega/giga (complex) LNG project life-cycles. Kozak (2004; p 2) states, "A benchmarking method consists of two parties: benchmarker and benchmarkee." Consequently, benchmarking Black Swan events involves evaluating and preparing for rare, high-impact occurrences that are difficult to predict but can have significant consequences.

6.10 What Are Black Swans

A Black Swan event is a term popularised by Nassim Nicholas Taleb to describe a rare, unexpected event that has a massive impact. These events are typically outside the realm of normal expectations and are only explainable in hindsight. In both small and mega projects, Black Swan events highlight the importance of being prepared for the unexpected. The existence of Black Swans was inconceivable as people had only seen White Swans in the northern hemisphere. Schauer (2003) points out that previously it was thought that all swans were white; however, the possibility remained open that there could be Black Swans. It was not until the discovery of Black Swans in Australia that this myth was refuted. Kendrick (2008) suggests that the realisation that Black Swans do exist gave rise to the metaphorical

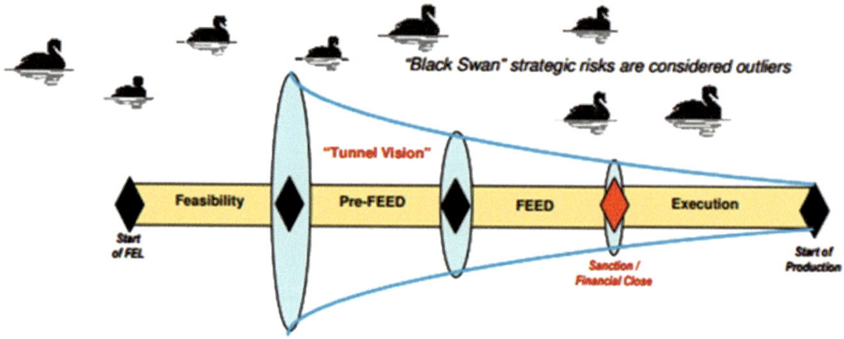

Figure 6.6 Black Swan Strategic Risks

Westney (2008; p 7)

use of the term Black Swans in projects, as an event that was believed to have no possibility of occurring. Consequently, Black Swans in projects are referred to as outliers, meaning the probability of the circumstance existing has not been predicted in the front end life-cycle of the project, refer to Figure 6.6. Westney (2008; p 7) states, ". . . strategic risks outside the assumptions underlying the base case are typically not addressed by the stage gate process. . . . black swan risks are typically not correlated with time or process . . ."

Therefore, when a Black Swan event does occur, usually in the execute phase of the project, it can have a catastrophic effect on the project outcome. Taleb (2004) suggests that, as Black Swans are unpredictable, instead of naively trying to predict them, we need to adjust our thinking to the reality of their existence. During the early phases of a project, when the one-off scenario is detected or is proposed as a random event, this can be identified as the early Black Swan syndrome. These findings are usually set aside during the Assess phase and not raised again. Consequently, by the time the Black Swan predictability becomes realistic, the previous concerns are not communicated due to fear of ridicule.

It is proposed, going forward, that the method of predicting Black Swans using conventional project management techniques needs to change within the project life-cycle timeframe to a more holistic view of how these low-probability, high-impact events occur. It would be projected that, during the early phases of a project life-cycle when Black Swans are identified, their potential presence is not disregarded as an impossibility but made clear to the JVPs to enable them to decide whether there is a need to set-up contingency funding to counteract them when/if they occur. This would be a below-the-line amount that would be carried through each phase of the project life-cycle and, should nothing occur after the execution phase, this allowance should be returned to the JVPs.

Presently, there is another phenomenon which is more extreme than Black Swans in projects. This extreme event is described by the author, as a single Red Swan, as illustrated in Figure 6.8. The author's definition of a Red Swan is an occurrence

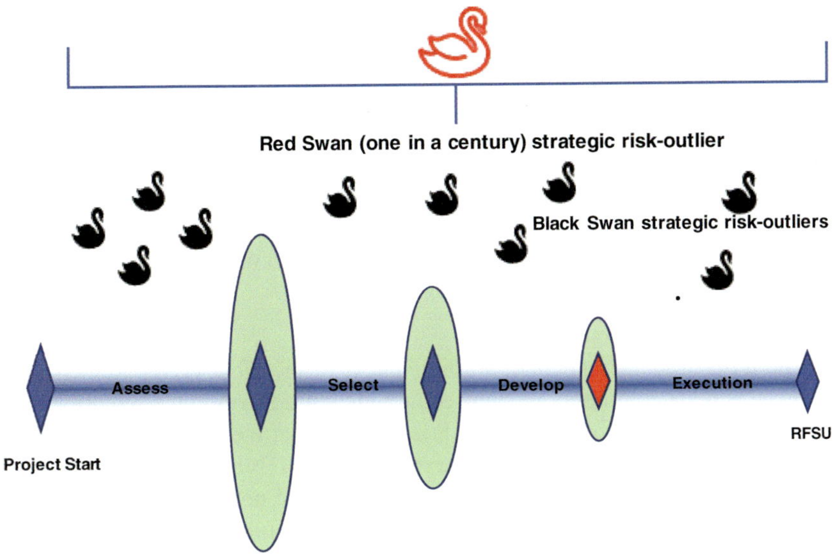

Figure 6.7 Identifies the Introduction of a Red Swan

Authors' Research

which is such a severe outlier, the probability of the existence is so inconceivable and hasn't happened for over a century. This extreme event influences not only one phase of an individual project but also all phases and indeed all projects worldwide. Recently, such a phenomenon occurred, of course, the COVID-19 pandemic. Kilbourne (2006) observed that there have been three major worldwide pandemics in the last century, viz:

- 1968 Hong Kong Influenza (H3N2);
- 1957 Asian Influenza (H2N2); and
- 1947 Pseudopandemic (H1N1).

Accordingly, the author notes that, including COVID-19, there have been four major worldwide pandemics within the last century. It is important to note that the term "pandemic" is typically used to describe the worldwide spread of a new infectious disease that affects many people. While there have been several significant outbreaks of infectious diseases in history, not all of them are categorised as pandemics. Subsequently, the four major worldwide pandemics are:

- 2019 COVID-19:
  The COVID-19 pandemic began in late 2019 and was caused by the novel coronavirus SARS-CoV–2. It has had a profound global impact on public health, economies, and daily life.

- 1968 Hong Kong Influenza (H3N2):
  The H3N2 influenza pandemic, also known as the Hong Kong flu, emerged in 1968. It was caused by a strain of the influenza A virus and spread globally. While it did result in significant illness and death, it is not as well-remembered as some other pandemics.
- 1957 Asian Influenza (H2N2):
  The H2N2 influenza pandemic, also known as the Asian flu, occurred in 1957. It was another global influenza pandemic caused by a new strain of the influenza A virus.
- 1947 Pseudopandemic (H1N1).
  The term "Pseudopandemic" in the context of the 1947 H1N1 influenza outbreak is not a commonly recognised or widely used term in the scientific literature or public health discussions. It is possible that the term is being used in a specific historical or informal context. The 1947 H1N1 influenza outbreak may have been a significant influenza event, but it did not achieve the status of a widely recognized worldwide pandemic like some other influenza outbreaks in history.

The recent COVID-19 pandemic had a profound and far-reaching impact on virtually every industry and projects around the world, including mega (complex) LNG projects, as well as smaller-scale and less complex ones. This is due, in part, to countries closing their borders, which affects international travel and materials/equipment deliveries to all projects, especially these multifaceted LNG projects in the Execute phase. As these projects are too large for any one entity to provide all the procurement, module fabrication, vessels, and design, the operations/processes are therefore dispersed across many countries. However, the lessons learnt during this period, such as the need for greater flexibility, digital transformation, and risk diversification, will likely influence how such projects are approached in the future.

6.11 Chapter Summary

In order for these mega/giga (complex) LNG projects to be delivered on schedule and within the budget, it requires a significant shift in organisational approach. By embracing changes and implementing best practices, organisations can enhance their capability to deliver mega LNG projects successfully, meeting both schedule and budget targets. It could be argued that organisations ought to train managers/personnel to deliver these multifaceted projects on time and budget, which is a challenge as this takes time, money, and dedication from senior management to support such a scheme.

This chapter highlights the importance of project life-cycle performance measurements, such as CIs and benchmarking, in enhancing the management and outcomes of future mega/giga (complex) LNG projects. Leveraging performance measurements such as these throughout the life-cycle of mega/giga LNG projects enhances project management effectiveness, ensures alignment with strategic objectives, and contributes to achieving successful outcomes. The theoretical process of utilising such performance measurements in the future life-cycles of mega/giga LNG projects aims to enhance transparency and efficiency across the entire project life-cycle model. Integrating Critical Indicators (CIs) and benchmarking into the theoretical framework of mega/giga LNG projects promotes transparency, efficiency, risk management, continuous improvement, and stakeholder accountability throughout the project life-cycle. This allows project managers to ensure that projects are developed and managed with clarity, effectiveness, and alignment with strategic goals, ultimately leading to successful outcomes and sustainable operations.

This has illustrated what constitutes a small project compared to mega or giga projects, together with the multifaceted nature of these large LNG projects as opposed to the traditional small LNG projects. It highlights the difficulty of developing a benchmarking strategy for these large projects compared to benchmarking traditional (small) projects and the dilemma in finding the path forward. In essence, developing a benchmarking strategy for large projects requires careful consideration of their unique characteristics and challenges, aiming for robust metrics that provide meaningful insights into project progress and performance. Key considerations throughout the life-cycle of mega LNG projects include regulatory compliance, environmental impact assessments, safety standards, supply chain management, and stakeholder engagement. Understanding these distinctions is crucial for successfully navigating the challenges associated with large-scale LNG projects. Each phase requires careful planning, execution, and monitoring to ensure project success and long-term sustainability.

7 Conclusion

7.0 Overview

Liquefied natural gas (LNG) is a crucial player in the global energy landscape. From the early stages of conception to the eventual decommissioning, the life-cycle of an LNG project is a testament to human ingenuity, determination, and cooperation, which can shape complex projects that contribute significantly to global energy supply while advancing sustainable development goals. Rahmat et al. (2022; p 1) state, "In the race towards greener energy, Liquified Natural Gas (LNG) stands out as an excellent alternative to reduce greenhouse gas emissions and help combat global warming." However, it is important to note that LNG is still a fossil fuel and not a zero-emission energy source. To effectively combat global warming, a comprehensive approach is needed that includes increasing renewable energy deployment, energy efficiency improvements, and efforts to reduce overall fossil fuel consumption and emissions.

As an LNG project progresses through its phases, the author's discussions have unveiled various facets of this intricate industry. Delving into the intricacies of benchmarking reveals the meticulous art of comparing and improving. The project's life-cycle was examined, from its feasibility and construction to its operation phase and eventual project closure.

The world's energy needs continue to evolve, driving innovation in LNG production, transportation, and utilisation. The pursuit of cleaner, more sustainable energy sources challenges organisations to reconsider traditional practices and embrace new technologies. Driving a paradigm shift in organisational strategies, technologies, and operational practices. Embracing changes not only addresses environmental challenges but also fosters innovation, resilience, and long-term competitiveness in the evolving energy landscape.

The future of LNG projects has a positive impact and uncertainty in equal measure. According to the author's observations and experience, while the future of LNG projects appears promising in meeting global energy demands and supporting a transition, navigating uncertainties requires strategic planning, innovation, and proactive engagement with diverse stakeholders. Balancing these factors will be key to realising the potential of LNG projects in a sustainable and competitive energy landscape. Climate change concerns, market fluctuations, and geopolitical

DOI: 10.1201/9781003688334-7

shifts will undoubtedly influence the path ahead. However, adaptability, resilience, and a commitment to responsible practices are crucial for the LNG industry to maintain a pivotal role in meeting the world's energy demands. By embracing adaptability, resilience, and responsible practices, the LNG industry can navigate challenges and capitalise on opportunities to meet the world's growing energy demands sustainably. Continual innovation and collaboration across sectors will be essential to ensure LNG remains a vital component of the global energy mix while advancing towards a more sustainable energy future.

7.1 Chapter Introduction

The aim of this book has been to develop a methodology for benchmarking mega/giga complex LNG projects throughout their project life-cycle, as well as providing a framework for future methods of benchmarking LNG projects. The author's observations and experience have contributed to a deeper understanding and management of mega/giga LNG projects by creating a methodology for benchmarking their performance and by offering a framework that can be used for benchmarking LNG projects in the future. This approach can enrich theories and practices by providing real-world insights, illustrating concepts through examples, and highlighting nuances that might not be captured in purely theoretical frameworks. Consequently, this can be valuable for stakeholders involved in the planning, execution, and oversight of such large-scale and complex energy projects.

This chapter includes the conclusions and recommendations from the author's investigations in the preceding chapters and how this contributes to the benchmarking of mega/giga complex LNG projects throughout their life-cycle and their phase durations. The author's unique assessment provides the framework on how these projects are benchmarked and evaluates how organisations benchmark their mega/giga complex LNG projects, and subsequent discussions with experienced project managers helped to validate the findings.

The conclusions from this book have confirmed and validated the hypothesis in performing this research on benchmarking mega/giga complex LNG projects. The author's observations highlight a persistent issue in multifaceted LNG projects: despite being thoroughly benchmarked, they often exceed their budgets and schedules. This suggests that while best practices and standards are in place, various factors, such as project complexity, regulatory challenges, or unforeseen circumstances, continue to derail timelines and cost management. Addressing these challenges may require a more nuanced approach to project planning and execution. This leads to query what data organisations are using to benchmark these multidimensional mega/giga LNG projects, as there are only a few that have been completed and all of which have had large schedule creeps and budget overruns. The author's evaluation has demonstrated that there is a gap in the current literature for benchmarking these mega/giga complex LNG project life-cycle phase durations. This investigation aims to address this shortcoming, culminating in the finalisation of this conclusion.

The insights gained from extensive experience in developing mega/giga complex LNG projects throughout their life-cycle, along with feedback from industry professionals through interviews, questionnaires, and informal discussions, highlighted several key factors crucial for the execution of these complex undertakings. The author underscores the importance of a holistic approach to developing complex LNG projects, emphasising that success hinges on careful planning, effective collaboration, and adaptability to changing circumstances. This data collection method enabled a clearer understanding of how organisations benchmark these mega complex LNG projects when only a few have been developed. As aforementioned, unfortunately, this benchmarking process is only performed in the Assess (initiate) phase to establish the roadmap for each phase to the completion of the Execute phase to obtain RFSU. As such, some organisations believe there is no need to revisit the benchmarking as the project proceeds through the other phases.

This book demonstrates that oil and gas organisations do benchmark the phases of the project life-cycle using benchmarking societies, etc. However, as aforementioned, only a few of these multifaceted mega complex LNG projects have been completed; therefore, the data the organisations use for benchmarking these LNG projects is unsuitable for complex mega/giga LNG projects as they have had large schedule creeps and cost overruns in the Execute phase. Observations suggest that this could be avoided if they revisited the duration for the Execute phase during the Develop phase. To do so, however, would potentially move the RFSU to the right, which would be difficult to convince the stakeholders to agree to. Beyond the development of a specific methodology, the book aims to establish a broader framework for benchmarking mega complex LNG projects in the future. This framework includes guidelines, key performance indicators (KPIs), data collection methods, and analysis techniques that can be applied not only to the projects discussed in the book but also to other LNG projects that may arise in the future. By providing such a framework, the book can contribute to the ongoing improvement of project management practices in the LNG industry.

It is recommended that oil and gas organisations should be approached to discuss and consider improved methods of benchmarking these multifaceted mega/giga complex LNG projects. The contention is how to implement a culture of change in organisations that maintain entrenched beliefs that their current methods and procedures of benchmarking mega/giga LNG projects is still fundamentally correct, regardless of the continual overruns on schedule, budgets, and scope creeps.

The framework is a guideline for the durations of each phase of the project life-cycle and a supplementary framework of the essential activities within each phase. Consequently, this highlights the importance of not fast-tracking the Develop phase to obtain FID and the pitfalls in doing so. Particularly, as mentioned in Chapters 1 and 4, there have been restrictions in many countries due to the COVID-19 virus, which has impacts on schedules and budgets.

These types of restrictions have significantly impacted supply chain distributions, labour shortages, and productivity in module yards. As noted in Chapter 3, this phenomenon is referred to as an outlier, which the author's has characterised as a 'Red Swan'. This outlier is so extreme that no-one could have predicted it because such an outlier is recognised as a one-in-a-hundred-year event. This is in contrast with the conventional term for outliers, i.e., 'Black Swans' which potentially could be predicted.

This book has provided an overview of the ramifications of the COVID-19 pandemic, and how companies now deal with this extremely rare event that has effects on all projects worldwide. The presence of this 'Red Swan' has meant that companies have had to reassess and re-evaluate their approach in developing mega/giga complex LNG projects. Most personnel and the fabrication of these mega/giga complex LNG projects are located in overseas locations, such as fabrication yards, design houses, and manufacturing companies.

This situation has necessitated an innovative approach in dealing with the restrictions whereby staff are confined to travel back to fabrication yards and design offices, etc. Personnel circumvent this now working from home where video conferencing, such as WebEx, is used to manage the development of these mega/giga complex LNG projects which, in turn, adds a further layer of complexity into the equation.

7.2 Research Strategy/Approaches

The research included interviews (questionnaires), survey questionnaires with participants that have worked on mega/giga complex LNG projects in Australia and worldwide. The research strategy in this book provides a sequential triangulation using qualitative and quantitative methodologies with the exploratory studies, interviews, and a case study. Tonon 2015; p 55 states, "Sequential Triangulation implies the use of quantitative and qualitative methods, with a limited interaction between both methods during the stage of data gathering and complementing each other's results at the moment of their interpretation."

The scrutiny approach determined that there is a need to identify common indicators for complex mega/giga LNG projects, which includes what measurement criteria are to be used to develop mega/giga complex LNG projects throughout their life-cycle. The author's observations have also found that the traditional constraints of the iron triangle are no longer suitable for the measurement performance on large LNG projects. The author has also determined that tools need to be developed that enhance the iron triangle by identifying the inner and outer triangles to include EVMS, KPIs, and the BSCs approach.

This type of process is the triangulation for the way performance measurements are structured, reference to Chapter 2 Figure 2.2. By doing so, this aids data capture by involving multiple research methods or measures.

In the early phases of the author's investigation, the data collection provided a performance measurement for the optimal means of how to improve benchmarking for success when developing these mega/giga complex LNG projects.

The structure of the investigation has considered all aspects of managing these complex LNG projects, including how/if tacit knowledge/information is good enough to be used as a strategy for changing benchmarking durations or whether it is considered unacceptable to do so. It should be noted that the interpretivist approach discussed in Chapter 3 includes the ideas (tacit information) that individuals bring into a project. This tacit information should not be ignored but should be reviewed to ascertain its viability to the success of these massive LNG projects.

The author's examination of mega complex LNG projects included the use of a combination of positivist and interpretivist approaches that were undertaken due to the multifaceted environment associated with mega/giga complex LNG projects, and the worldwide review that was part of the investigation for the development of these complex LNG projects. It was imperative that, to measure complex projects, a deep understanding is required of their structure and the human interactions involved. It is not just about the technical aspects or the raw data; it is about comprehending how different components interact, how teams collaborate, and how decisions are made throughout the project life-cycle. This understanding is crucial for accurately assessing progress, identifying potential risks, and ultimately ensuring the project's success.

The author's observation is that these complex projects are from human endeavours and scientific approaches alone are not adequate to measure them; therefore, it was deduced that there must be a combination of both scientific and human endeavours to provide a performance measurement. Using the triangulation approach enabled the capture of all aspects of data to establish a performance measurement for these complex LNG projects.

The author's guidelines in this book for the structure of what the deliverables are for each phase of a mega/giga complex LNG project. This also suggests that mid-stream review gates need to be considered for each phase of the development of these multifaceted LNG projects.

The author's observations have found that, due to the multidimensional structures of these multifaceted LNG projects, it is difficult for many organisations to comprehend or predict the complexity of the SoW whilst they are in the early phases of the project life-cycle.

7.3 Exploratory Study

The exploratory study examines the life-cycle performance measurements and how these insights are interpreted in this book. Exploratory studies are invaluable for delving into the nuances of life-cycle performance measurements within the context of the book. They offer an opportunity to uncover hidden patterns, identify emerging trends, and gain a deeper understanding of how these measurements are interpreted in practice. By examining real-world examples and engaging with stakeholders, such studies can shed light on the complexities and challenges inherent in interpreting performance metrics throughout the project life-cycle. This insight is then distilled and presented in the book, providing readers with a comprehensive understanding of how to effectively measure and interpret performance in complex projects.

Consequently, empirical studies provide conclusions and insights into the project complexity of the phases in these multifaceted LNG projects. Exploratory studies for project management research provide insights into the future project risks of these complex LNG projects.

7.4 Future of Mega LNG Projects (The Recent COVID-19 Pandemic)

What does the future hold for organisations developing mega/giga complex LNG projects that faced a range of challenges during the COVID 19 pandemic. In order to comprehend this, one must appreciate that these mega/giga complex LNG projects are also modularised. These modules are manufactured/fabricated in international locations, such as China, South Korea, etc., and then shipped to their LNG construction sites around the world for installation on the LNG plots. The author's observation has been that while organisations developing mega or giga LNG projects faced significant challenges during the COVID-19 pandemic, the future holds opportunities driven by global energy demand recovery, sustainability goals, technological advancements, and evolving geopolitical dynamics. Successfully navigating these factors will be crucial to the sustainability and profitability of LNG projects in the post-pandemic era. Organisations success will depend on their ability to innovate, manage risks effectively, and align with evolving global energy trends.

7.5 Recommendations for Managing Mega/Giga LNG Projects

From the author's observations and experience in these multifaceted projects, it is evident that there needs to be changes in the management style when developing these mega/giga complex LNG projects.

7.5.1 Recommendations to Prevent Schedule and Cost Overruns

From the author's observations and experience, it is recommended that, to help prevent the continuous schedule and cost overruns, benchmarking changes should be made when the organisation begins to develop these mega/giga complex LNG projects from the onset. These changes should include, but not be limited to, the following:

- how the organisation is managing the development of the overall project life-cycle (visibility on decision-making at the top level);
- how there is a requirement to align the corporate and project governance teams that is visible to all within the project team;
- project phase durations of the Develop and Execute phase should be reviewed, without bias, to reach any stretch targets for FID or RFSU;
- the construction (Execute) management team (not just one or two individuals) should be involved at the start of the Develop phase;
- selection criteria for selecting the project director(s) and the project manager(s) should have more clarity, and their competencies reviewed; and
- training of a new category of project manager/director for mega/giga complex LNG projects should be considered.

7.6 Future Recommendations

There is a considerable gap in the existing literature regarding the life-cycle phase durations of mega/giga LNG projects. Despite the growing importance of these complex undertakings in the global landscape, there is a scarcity of comprehensive studies that analyse the timelines associated with various phases of these projects, such as planning, construction, commissioning, and operation. As the demand for LNG continues to rise, addressing this disparity in literature becomes increasingly critical. Future research efforts could focus on collecting empirical data from completed projects, conducting case studies, and developing standardised frameworks to analyse and compare the life-cycle phase durations of mega and giga LNG ventures.

Furthermore, additional evaluations are required to establish and/or validate the life-cycle durations proposed in this book.

When developing mega/giga complex LNG projects in the oil and gas industry, prioritising the Develop phase in mega/giga LNG projects is not just a recommendation but a critical necessity. It ensures that the complexities inherent in such projects are thoroughly understood, risks are mitigated, stakeholders are aligned, resources are effectively managed, and the project is set up for successful execution and long-term viability. This approach enhances the overall project governance, reduces uncertainties, and increases the likelihood of achieving project objectives within established timelines and budgets.

The development phase of mega/giga LNG projects ensures that the SoW of these multifaceted projects is completely visible. This would facilitate the stakeholders and the team members to understand and appreciate the complexity of developing these multidimensional mega/giga LNG initiatives. The contention here is that there is a great deal of peer pressure from the client stakeholders on the teams to achieve the Financial Investment Decision (FID) to move into the Execute phase to enable them to achieve their gas contract delivery dates post-RFSU.

The author's analysis combines theoretical insights with practical experience in developing mega/giga LNG projects, it becomes highly beneficial for project managers and their teams. The author has also provided a template of the minimal key deliverable requirements that are necessary within each phase of the mega/giga project life-cycle. As there appears to be a significant discrepancy in knowledge between the current and previous project professionals in the oil and gas industry, the template provided represents a pragmatic means of undertaking these multifaceted mega/giga complex LNG projects. The concept is to re-appraise and focus more on the practical means of developing these LNG projects, such as a more hands-on approach and not merely theoretical.

For future mega/giga complex LNG projects, the issue is how to introduce the recommended modifications to the benchmarking durations proposed in this book for the life-cycle phases. Organisations utilise the benchmarking houses, which are reliable benchmarking organisations, to provide accurate data. However, it could be argued that the organisations do not possess hands-on experience in developing these complex LNG projects but instead were only gathering data. Correspondingly, it must be acknowledged that the information provided to the benchmarking organisations is that of the status at the assess (initiate) phase. Consequently, this renders the initial SoW outdated and irrelevant as the information initially provided during the Assess (initiate) phase becomes obsolete due to the lack of the current status or needs of the project. Therefore, organisations typically do not continue to engage in benchmarking organisations using the original SoW because it does not accurately represent the project's current state or objectives.

The primary objective of the book is to create a structured and systematic approach for evaluating large and complex LNG projects. Benchmarking is a structured process of comparing and evaluating the performance, processes, or outcomes of one project (or organisation) against others that are similar or complete. This practice assists in identifying the best practices that lead to superior performance, areas where improvements can be implemented, and opportunities to optimise project management and execution. In the context of mega/giga LNG projects, benchmarking can be advantageous in ensuring that these high-value and complex endeavours are executed efficiently and effectively.

7.7 Outcomes

Based on observations and experience, it appears there is a gap in the literature when it comes to pragmatic guidance specifically tailored for benchmarking durations and minimum requirements across phases for mega/giga LNG projects. Although there is extensive literature on benchmarking mega projects, much of it concentrates on infrastructure projects, such as transportation, construction, and utilities. The author's research has determined that there is a notable gap when it comes to pragmatic knowledge specifically addressing the benchmarking durations and minimum requirements for each phase of mega/giga LNG projects. Although there is significant literature on benchmarking mega projects, the specific needs and complexities of mega/giga LNG projects necessitate a focused effort to develop pragmatic knowledge and theoretical frameworks tailored to the unique characteristics of the LNG industry. Addressing this issue will enhance the efficiency and effectiveness of LNG project management globally.

Organisations developing these multifaceted LNG projects need to make an ideological change, which ought to take place within organisations that continue to use traditional project management techniques and processes to develop mega/giga complex LNG projects, as these are not robust enough to deal with the complexities associated with these endeavours. This is suggested by many reputable authors like Merrow and Flyvbjerg, who have been vocal about the limitations of traditional project management processes when applied to mega LNG projects. Merrow and Flyvbjerg reflect a growing recognition within the industry that traditional project management approaches are often inadequate for mega projects. There is a clear need for organisations to embrace integrated, adaptive, and agile project management practices that can effectively address the diverse challenges of these large-scale endeavours. By doing so, organisations can enhance their ability to deliver mega projects on time, within budget, and in a sustainable manner, thereby maximising their success and minimising risks.

The concern is that it would be difficult to reform within organisations developing mega LNG projects, which is the strategy of using stretch targets on these massive projects to obtain FID and the RFSU dates. These types of strategies only compound the issues of schedule and cost overruns.

Furthermore, the author's observation is that many experienced team members are fully aware that these stretch targets, normally set by the organisation's executives, are unrealistic resulting in the team members becoming discouraged at the prospect of failure.

7.8 Theoretical and Practical Contributions

This book has provided a feasible and practical contribution to the basis of measuring the durations of the life-cycle phases of mega/giga complex LNG projects. By providing a structured approach to measuring and benchmarking durations across the life-cycle phases of mega/giga LNG projects, the book supports project teams in achieving efficiency, cost effectiveness, and ultimately, successful project delivery. It is a valuable resource for stakeholders involved in planning, executing, and managing LNG projects, resulting in improved project management practices and sustainable development outcomes. Subsequently, by using the tables and figures within this book, which can be adjusted, according to the size and complexity of the project that the organisations are undertaking, this can assist in the project's success.

Prior to this assessment, there has been limited theoretical or hypothetical contributions on the life-cycle phases and the individual phase durations of these complex LNG projects. Therefore, this data provided by the author's observations can contribute to the sustainability of developing these multifaceted LNG projects by providing a practical benchmarking framework as a guideline. The fusion of theory and practice in this context is invaluable, as it empowers industry professionals, researchers, and decision-makers with the tools and insights needed to navigate the intricate world of LNG projects development effectively.

7.9 Chapter Summary

Based on the author's observation and experience in developing mega/giga complex LNG projects throughout their life-cycle, there is a great deal that organisations and project teams ought to consider, including sustainability, when benchmarking these multifaceted LNG projects for success. By integrating sustainability into the benchmarking process, organisations and project teams can foster responsible development practices, enhance stakeholder trust, and ensure the long-term success and sustainability of mega/giga LNG projects. The issue is how to get organisations to adhere to sustainability standards for developing these multifaceted LNG projects. In this context, this book has provided a practical benchmarking framework as a guideline in Chapter 4 – section 4.3. This can be a practical guideline for organisations to adopt sustainable practices in their LNG projects development. Utilising benchmarks can help organisations compare their current practices with industry standards or best practices, identify areas for improvement, and set goals for sustainability.

From the author's interpretation, this book recognises the importance of organisational changes to address challenges such as schedule and cost overruns, as well as scope creeps, in future mega/giga complex LNG developments. These types of projects are indeed complex and often involve significant investments of time, resources, and capital. Managing them effectively requires not only technical expertise but also robust organisational structures and processes. Addressing issues like schedule and cost overruns and scope creeps often involves improving project management practices, enhancing communication and collaboration among stakeholders, implementing effective risk management strategies, and ensuring alignment between project objectives and organisational goals.

The transition to new benchmarking and management structures in industries such as LNG, which involve complex projects and substantial investments, presents a variety of challenges. Challenges require a strategic approach that considers the specific dynamics of the LNG industry. It involves thorough planning, stakeholder engagement, risk mitigation strategies, and leveraging technology to drive operational excellence and sustainable growth. Taking a step back from traditional project management life-cycle processes for mega/giga LNG projects requires a concerted effort and determined organisational support/assistance. CEOs play a crucial role in leading organisational change processes. As a champion of change, they can provide the vision, direction, and support necessary to drive the adoption of new frameworks and practices. Changing organisations' benchmarking and management structures can be an arduous process to undertake. However, achieving a departure from the current traditional project management life-cycle processes for mega/giga LNG phase durations requires determined organisational support to accept the proposed life-cycle framework for phase durations. This is a long and complicated process which would need to be championed by the CEOs of organisations to lead the management of change process to achieve this and enhance the benchmarking durations being recommended in this book.

Setting realistic and attainable targets is crucial for the success of complex projects like LNG ventures. Insisting on stretch targets that are beyond feasibility can lead to detrimental consequences such as cost overruns, delays, and compromised quality. It is essential for organisations, corporate management, and joint venture partners to recognise the complexity of these projects and align their objectives accordingly. By establishing achievable objectives, stakeholders can promote sustainable progress and minimise risks, ultimately resulting in the long-term success of the project.

To try to accelerate the FID approval and RFSU end dates of these complex projects only adds to the project failure with continued cost and schedule overruns and potential scope creeps. Attempting to accelerate the approval and execution dates of complex projects like LNG ventures without due consideration of their intricacies can indeed exacerbate the risk of failure. Rushing through critical decision-making processes, such as Final Investment Decision (FID) approval and Ready for Start-Up (RFSU) dates, often leads to overlooking important details, inadequate planning, and increased chances of cost and schedule overruns. Ultimately, a balanced approach that considers both speed and thoroughness is crucial for the success of complex LNG projects.

To ensure project success, it is important that the project director of these mega/giga LNG projects continues with the project from the Develop phase to the end of the Execute phase.

The author's recommendation is that further research should be undertaken with the oil and gas organisations to ascertain if benchmarking these changes to the proposed life-cycle phase durations mentioned in this book would be considered when undertaking the mega/giga complex LNG projects.

Following this chapter, the author's observations reveal that knowledge is a beacon guiding us forward. It is a collective endeavour fuelled by curiosity and a thirst for progress. Whether you are a member of the LNG industry, a curious observer, or an advocate for sustainability, your role in shaping the future is significant.

Let this conclusion be a reminder that our understanding of LNG projects, such as the industry itself, remains dynamic and ever-evolving. The journey doesn't end here; it merely paves the way for new beginnings and discoveries.

References

Abdel-Aal, H.K., ed. 2021. *Economic Analysis of Oil and Gas Engineering Operations*. Boca Raton: CRC Press. https://doi.org/10.1201/9781003137696.

Akampurira, A. 2014. *Project Planning and Management: An Aspect of Development*. Hamburg, Germany: Anchor Academic Publishing.

Akintoye, A. 2010. "Analysis of Factors Influencing Project Cost Estimating Practices." *Construction Management and Economics* 18 (1): 77–89. https://doi.org/10.1080/014461900370979.

Anand, G., and R. Kodali. 2008. "Benchmarking the Benchmarking Models." *Benchmarking: An International Journal* 15 (3): 257–291. https://doi.org/10.1108/14635770810876593.

Andersen, B., and P. Pettersen. 1996. *The Benchmarking Handbook*. London: Chapman & Hall.

Andersen, B., K. Samset, and M. Welde. 2016. "Low Estimates – High Stakes: Underestimation of Costs at the Front-End of Projects." *International Journal of Managing Projects in Business* 9 (1): 171–193. https://doi.org/10.1108/IJMPB-01-2015-0008.

Andersen, E.S., A. Dysvik, and A.L. Vaagaasar. 2009. "Organizational Rationality and Project Management." *International Journal of Managing Projects in Business* 2 (4): 479–498. http://dx.doi.org/10.1108/17538370910991106.

Arena, L., and E. Molloy. 2010. "The Governance Paradox in Megaprojects." *Entretiens Jaques Cartier* 1 (1). Lyon, France. https://www.academia.edu/49107002/The_Governance_Paradox_in_Megaprojects.

Aritua, B., N.J. Smith, and D. Bower. 2009. "Construction Client Multi-Projects – A Complex Adaptive Systems Perspective." *International Journal of Project Management* 27 (1): 72–79. Leeds, UK: School of Civil Engineering, University of Leeds. https://doi.org/10.1016/j.ijproman.2008.02.005.

Arndt, J., and S. Edrington. 2011. *Rural and Urban Transit District Benchmarking – Effectiveness and Efficiency Guidance Document*. Texas: Texas Transportation Institute. https://rosap.ntl.bts.gov/view/dot/20943.

Ashkanani, S., and R. Franzoi. 2022. "An Overview on Megaproject Management Systems." *Management Matters* 19 (2): 129–148. https://doi.org/10.1108/MANM-01-2022-0006.

Australian Bureau of Statistics. January 2022. *International Trade in Goods and Services, Australia*. https://www.abs.gov.au/statistics/economy/international-trade/international-trade-goods/jan–2022.

Australian Government Department of Industry, Science and Resources. December 2023. *Resources and Energy Quarterly*. https://www.industry.gov.au/sites/default/files/2023-12/resources-and-energy-quarterly-december–2023.pdf.

Awwad, M.A., A.M.A. Radhi, M.K., Panigrahy, and S.K. Gopal. 2021. "Cost Optimisation in Mega Oil & Gas Projects." Paper Presented at the Abu Dhabi International Petroleum Exhibition and Conference, Abu Dhabi, UAE. https://doi.org/10.2118/207751-MS.

Baek, C., and J. Lee. 2009. “The Relevance of DEA Benchmarking Information and the Least-Distance Measure.” *Mathematical and Computer Modelling* 49 (1–2): 265–275. https://doi.org/10.1016/j.mcm.2008.08.007.

Bogle, J.C. 11 October 2007. “ Black Monday and Black Swans.” Paper Presentation at the Risk Management Association, Boca Raton, Florida. https://boglecenter.net/wp-content/uploads/risk_mgmt.pdf.

Brandy, T., and A. Davies. 2014. “Management Structural and Dynamic Complexity: A Tale of Two Projects.” *Project Management Journal* 45 (4): 21–38. http://dx.doi.org/10.1002/pmj.21434.

Briggs, G., 2010. *The Global Energy Demand for Australian Sourced LNG*. Perth: Future Directions International Pty Limited.

Cameron, D., H.M. With, Ø. Skjerven, M. Hyllseth, and P. MannsåKer. 2005. “Integrated Distributed Dynamic Simulation of the LNG Value Chain: Concepts and Experience.” *Chemical Engineering Research and Design* 83 (6): 686–691. https://doi.org/10.1205/cherd.04372.

Cantoni, F., and F. Pagnone. 2020. “Megaprojects. A Special Eye on Sustainability to Overcome the “Iron Law”.” In Favari, E., and F. Cantoni, eds. *Megaproject Management a Multidisciplinary Approach to Embrace Complexity and Sustainability*, 1–14. Piacenza, Italy: SpringerBriefs in Applied Sciences and Technology.

Cassidy, N., and M. Kosev. 2015. “Australia and the Global LNG Market.” *Reserve Bank of Australia Bulletin March Quarter* 33–43. https://www.rba.gov.au/publications/bulletin/2015/mar/4.html.

Cooper, R.G. 2008. “Perspective: The Stage-Gate® Idea-to-Launch Process – Update, What’s *New*, and NexGen Systems.” *Journal of Product Innovation Management* 25 (3): 213–232. https://doi.org/10.1111/j.1540-5885.2008.00296.x.

Cortada, J.W., and J.A. Woods, eds. 2000. *The Knowledge Management Yearbook 2000–2001*. London: Taylor & Francis Group. https://doi.org/10.4324/9780080941042.

Criaud, M., and P. Boulinguez. 2013. “Have We Been Asking the Right Questions When *Assessing* Response Inhibition in Go/No-Go Tasks with fMRI? A Meta-Analysis and Critical Review.” *Neuroscience & Biobehavioral Reviews* 37 (1): 11–23. http://dx.doi.org/10.1016/j.neubiorev.2012.11.003.

Department of Industry, Science and Resources. July 2023. *Resources and Energy Quarterly: June 2023*. Office of the Chief Economist. https://www.industry.gov.au/publications/resources-and-energy-quarterly-june−2023.

Drennan, L.T., A. McConnell, and A. Stark. 2014. *Risk and Crisis Management in the Public Sector*. 2nd ed. Oxon: Routledge Taylor & Francis Group.

Eftimie, C. 2021. “Processing Equipment Tips: Consider Modularization.” *Chemical Industry Digest* 34. https://www.chemicalprocessing.com/articles/2017/processing-equipment-tips-consider-modularization/.

Egan, J. 1998. *Rethinking Construction: The Report of the Construction Task Force, DTI (URN 98/1095)*. UK: Construction Task Force.

Elmuti, D., and Y. Kathawala. 1997. “An Overview of the Benchmarking Process: A Tool for Continuous Improvement and Competitive Advantage.” *Benchmarking for Quality Management & Technology* 4 (4): 229–243. https://doi.org/10.1108/14635779710195087.

Fleming, Q.E., and J.M. Koppelman. 2010. *Earned Value Project Management*. 4th ed. Newton Square: PMI.

Flyvbjerg, B. 2014. “What you Should Know About Megaprojects and Why: An Overview.” *Project Management Journal* 45 (2): 6–19. https://doi.org/10.1002/pmj.21409.

Flyvbjerg, B., N. Bruzelius, and W. Rothengatter. 2003. *Megaprojects and Risk: An Anatomy of Ambition*. Cambridge: The Press Syndicate of the University of Cambridge University Press.

Fong, S.W., E.W.L. Cheng, and D.C.K. Ho. 1998. “Benchmarking: A General Reading for Management Practitioners.” *Management Decision* 36 (6): 407–418. https://doi.org/10.1108/00251749810223646.

Geoscience Australia. 2012. "Location of Australia's Gas Resources and Infrastructure." *Australian Gas Resource Assessment 2012*. https://www.ga.gov.au/bigobj/GA21116.pdf.

Geoscience Australia. Australia's Energy Commodity Resources. 2024. *Australia's Remaining Gas Reserves (2P) and Contingent Resources (2C) During 2022 and Cumulative Production to End 2022 (PJ)*. Gas | Geoscience Australia.

Giezen, M., L. Bertolini, and W. Salet. 2014. "Adaptive Capacity Within a Mega Project: A Case Study on Planning and Decision-Making in the Face of Complexity." *European Planning Studies* 23 (5): 999–1018. https://doi.org/10.1080/09654313.2014.916254.

Goodwin, D.L., and W.L. Goodwin. 1999. "Research Design and Methodology Section: Measurement Myths and Misconceptions." *Springer International Publishing* 14 (4): 408–427. https://oce.ovid.com/article/00012097-199912000-00004/HTML.

Habibullah, A., P. Lardi, and M. Passmore. 8–11 March 2009. "LNG Conceptual Design Strategies." Paper Presented at the 88th GPA Annual Convention, San Antonio, Texas. http://www.anave.es/images/seguridad/worleyparsons-lng_conceptual_design_strategies-nov_12.pdf.

Haidar, A., and R. Ellis. 4–7 November 2010. "Analysis and Improvement of Megaprojects Performance." Paper Presented at the Engineering Project Organizations Conference, South Lake Tahoe, California. https://authorzilla.com/71w7d/analysis-and-improvement-of-megaprojects-performance.html.

Herkenhoff, L. 2018. *Profile of the Oil and Gas Industry: Resources, Market Forces, Geopolitics, and Technology*. New York: Business Expert Press, LLC.

Horváthová, J., M. Mokrišová, and M. Vrábliková. 2021. "Benchmarking – A Way of Finding Risk Factors in Business Performance." *Journal of Risk and Financial Management* 14 (5): 1–17. https://doi.org/10.3390/jrfm14050221.

Hussein A., A. AlNouss, S. Aseel, M. Kucukvar, N.C. Onat, and T. Al-Ansari. 2022. "How Sustainable is Liquefied Natural Gas Supply Chain? An Integrated Life Cycle Sustainability Assessment Model." *Energy Conversion and Management: X* (15): 1–19. https://doi.org/10.1016/j.ecmx.2022.100246.

Institute for Energy Economics and Financial Analysis. 2023. "Global LNG Exports." https://ieefa.org/sites/default/files/2024-04/Global%20LNG%20Outlook%202024-2028_April%202024%20%28Final%29.pdf.

Invernizzi, D.C., G. Locatelli, and N.J. Brookes. 2018. "A Methodology Based on Benchmarking to Learn Across Megaprojects: The Case of Nuclear Decommissioning." *International Journal of Managing Projects in Business* 11 (1): 104–121. https://doi.org/10.1108/IJMPB-05-2017-0054.

Iyer, K. C, and P.S. Banerjee. 2015. "Measuring and Benchmarking Managerial Efficiency of Project Execution Schedule Performance." *International Journal of Project Management* 34 (2): 219–236. https://doi.org/10.1016/j.ijproman.2015.10.008.

Jensen, J.T. 2003. "The LNG Revolution." *The Energy Journal* 24 (2): 1–45. https://doi.org/10.5547/ISSN0195-6574-EJ-Vol24-No2–1.

Katebah, M.A., M.M. Hussein, A. Shazed, Z. Bouabidi, and E.I. Al-Musleh. October 2020. "Rigorous Simulation, Energy and Environmental Analysis of an Actual Baseload LNG Supply Chain." *Computers & Chemical Engineering* 141. https://doi.org/10.1016/j.compchemeng.2020.106993.

Kaufman, R. 2000. *Mega Planning: Practical Tools for Organisational Success*. Thousand Oaks: SAGE Publications, Inc.

Keegan, A., and J.R. Turner. 2000. "Managing Human Resources in the Project-Based Organization." In Turner, J.R., S.J. Simister, and D. Lock, eds. *The Gower Handbook of Project Management*. 3rd ed. Aldershot: Gower.

Keehley, P., and S.A. MacBride. 1997. "Can Benchmarking Best Practises Work for Government?" *Quality Progress* 30 (3): 75–80. https://www.proquest.com/docview/214520712/fulltextPDF/3F2F7F2705C34477PQ/1?accountid=10382.

Kendrick, T. 2008. “Avoiding Black Swans: Managing Risks Using the PERIL Database.” Paper Presented at the PMI® Global Congress, 2008 – North America, Denver, CO, Newton Square, PA, Project Management Institute. https://www.pmi.org/learning/library/project-experience-risk-black-swans-7000.

Kerzner, H. 2006. *Project Management: A Systems Approach to Planning, Scheduling, and Controlling*. 9th ed. NJ: John Wiley & Sons, Inc.

Kerzner, H. 2013. *Project Management: A Systems Approach to Planning, Scheduling, and Controlling*. 11th ed. Somerset: John Wiley & Sons, Inc.

Kerzner, H. 2014. *Project Management Best Practices: Achieving Global Excellence*. 3rd ed. NJ: John Wiley & Sons, Inc.

Kerzner, H. 2018. *Project Management Best Practices: Achieving Global Excellence*. 4th ed. NJ: John Wiley & Sons, Inc.

Kidnay, J.A., W.R. Parrish, and D.G. McCartney. 2011. *Fundamentals of Natural Gas Processing*. 2nd ed. Boca Raton: CRC Press.

Kilbourne, E.D. 2006. “Influenza Pandemics of the 20th Century.” *Emerging Infectious Diseases Journal* 12 (1): 9–14. https://dx.doi.org/10.3201%2Feid1201.051254.

Koc-Menard, S. 2009. “Flexible Work Options for Older Workers.” *Strategic HR Review* 8 (2): 31–36. https://doi.org/10.1108/14754390910937567.

Koskinen, K., P. Pihlanto, and H. Vanharanta. 2003. “Tacit Knowledge Acquisition and Sharing in a Project Work Context.” *International Journal of Project Management* 21 (4): 281–290. https://doi.org/10.1016/S0263-7863(02)00030–3.

Kozak, M. 2004. *Destination Benchmarking Concepts, Practices, and Operation*. Cambridge: CABI Publishing.

Kulikowski, K. 2020. “The Model of Evidence-Based Benchmarking: A More Robust Approach to Benchmarking.” *Benchmarking: An International Journal* 28 (2): 721–736. http://dx.doi.org/10.1108/BIJ-04-2020-0175.

Lankford, W.M. 2000. “Benchmarking: Understanding the Basics.” *The Coastal Business Journal* 1 (1): 57–62. https://digitalcommons.coastal.edu/cgi/viewcontent.cgi?article=1007&context=cbj.

Limbrick, O. 2009. *Almost All You Wanted to Know About Project Benchmarking*. Texas: Construction Industry Institute.

Love, P.E.D., D.J. Edwards, Z. Irani, and D.H.T. Walker. 2009. “Project Pathogens: The Anatomy of Omission Errors in Construction and Resource Engineering Project.” *IEEE Transactions on Engineering Management* 56 (3): 425–435. https://doi.org/10.1109/TEM.2008.927774.

Love, P.E.D., J. Lui, J. Matthews, C.P. Sing, and J. Smith. 2015. “Future Proofing PPPs: Life-Cycle Performance Measurement and Building Information.” *Automation in Construction* 56: 26–35. https://doi.org/10.1016/j.autcon.2015.04.008.

Love, P.E.D., X. Wang, M.C.P. Sing, and R.L.K. Tiong. 2013. “Determining the Probability of Project Cost Overruns.” *Journal of Construction Engineering and Management* 139 (3): 321–330. http://dx.doi.org/10.1061/(ASCE)CO.1943-7862.0000575.

Ma, L., and H. Fu. 2020. “Exploring the Influence of Project Complexity on the Mega Construction Project Success: A Qualitative Comparative Analysis (QAC) Method.” *Engineering, Construction and Architectural Management* 27 (9): 2429–2449. http://dx.doi.org/10.1108/ECAM-12-2019-0679.

Macneil, J., J. Testi, J. Cupples, and M. Rimmer. 1995. *Benchmarking Australia Linking Enterprises to World Best Practice*. Melbourne: Pitman Publishing Asia Pacific.

Magni, C.A., and A. Marchioni. 2020. “Average Rates of Return, Working Capital, and NPV-*Consistency* in Project Appraisal: A Sensitivity Analysis Approach.” *International Journal of Production Economics* 229: 1–15 https://doi.org/10.1016/j.ijpe.2020.107769.

Marrewijk, A., S.R. Clegg, T.S. Pitsis, and M. Veenswijk. 2008. “Managing Public–Private Megaprojects: Paradoxes, Complexity, and Project Design.” *International Journal of Project Management* 26 (6): 591–600. https://doi.org/10.1016/j.ijproman.2007.09.007.

Mavi, R.K., and C. Standing. September 2018. "Critical Success Factors of Sustainable Project Management in Construction: A Fuzzy DEMATEL-ANP Approach." *Journal of Cleaner Production* 194: 751–765. https://doi.org/10.1016/j.jclepro.2018.05.120.

Mbassegue, P., M. Gardoni, and Z. Tahboub. 2017. "The Knowledge Benchmarking Process Framework: A New Basis to Analyse Megaproject Challenges and Practices." *International Journal of Knowledge Management and Practices* 5 (2): 7–15. https://www.proquest.com/docview/2024116785?pq-origsite=primo&accountid=10382.

McCabe, S. 1998. *Quality Improvement Techniques in Construction.* New York: Addison Wesley Longman Limited.

McIntosh, S.A. 2018. *Keys to Successful Major Project Execution.* Littleton: Long International, Inc.

McKenna, M.G., H. Wilczynski, and D. VanderSchee. 2006. *Capital Project Execution in the Oil and Gas Industry. Increased Challenges, Increased Opportunities*. Virginia: Booz Allen Hamilton, Inc.

Merrow, E.W. 1 June 2003. "Mega-Field Developments Require Special Tactics, Risk Management." *Offshore Magazine*. https://www.offshore-mag.com/business-briefs/article/16755442/megafield-developments-require-special-tactics-risk-management.

Merrow, E.W. 2011. *Industrial Megaprojects: Concepts, Strategies, and Practices for Success*. Hoboken: John Wiley & Sons, Inc.

Merrow, E.W. 2012. "Oil and Gas Industry Megaprojects: Our Recent Track Record." *Oil and Gas Facilities* 1 (2): 38–42. https://doi.org/10.2118/153695-PA.

Meza, A., M. Koc, and M.S. Al-Sada. 2022. "Perspectives and Strategies for LNG Expansion in Qatar: A SWOT Analysis." In *College of Science and Engineering Doha, Qatar*, 1–15. Elsevier. https://doi.org/10.1016/j.resourpol.2022.102633.

Miles, M.B., and A.M. Huberman. 1994. *Qualitative Data Analysis: An Expanded Sourcebook.* Thousand Oaks: SAGE Publications Ltd.

Mokhatab, S., J.Y. Mak, J.V. Valappil, and D.A. Wood. 2014. *Handbook of Liquefied Natural Gas.* Oxford: Gulf Professional Publishing.

Mokhatab, S., W. Poe, and J. Speight. 2006. *Handbook of Natural Gas Transmission and Processing*. Burlington: Gulf Professional Publishing.

Mukherjee, A., and P. Chatterjee. 2015. "A Framework for Understanding and Modeling Risk in Mega-Projects and Its Impact on the Markets for Project Finance." *Journal of Structured Finance* 21 (3): 88–105. http://dx.doi.org/10.3905/jsf.2015.21.3.088.

Nalewaik, A., and A. Mills. July 2015. "Project Performance Audit: Enhanced Protocols for Triple Bottom Line Results." *Procedia – Social and Behavioural Sciences* 194: 134–145. https://doi.org/10.1016/j.sbspro.2015.06.185.

Nooteboom, U. 2004. "Interface Management Improves On-Time, On-Budget Delivery of Megaprojects." *Journal of Petroleum Technology* 56 (8): 32–34. https://doi.org/10.2118/0804-0032-JPT.

Oakland, J.S. 2007. *Total Quality Management Text With Cases*. Oxford: Butterworth Heinemann.

Olaniran, G., P.E.D. Love, D.J. Edwards, O.A. Olatunji, and J. Matthews. 2015. "Chaotic Dynamics of Cost Overruns in Oil and Gas Megaprojects: A Review." *International Journal of Civil, Environmental, Structural, Construction and Architectural Engineering* 9 (7): 991–917. http://dx.doi.org/10.13140/RG.2.1.4628.0403.

Olugbenga, J.O., P.E.D. Love, D.J. Edwards, O.A. Olatunji, and J. Matthews. 2015. "Cost Overruns in Hydrocarbon Megaprojects: A Critical Review and Implications for Research." *Project Management Journal* 46 (6): 126–138. https://doi.org/10.1002/pmj.21556.

Patanakul, P., and A.J. Shenhar. 2012. "What Project Strategy Really is: The fundamental Building Block in Strategic Project Management." *Project Management Journal* 43 (1): 4–20. https://doi.org/10.1002/pmj.20282.

Patty, R.M., and M.A. Denton. 2010. *The End of Project Overruns: Lean and Beyond for Engineering, Procurement and Construction.* Boca Raton: Universal Publishers.

Phalen, T., and J. Scotti. 2009. "What is the Outlook for Building New LNG Facilities?" *Hydrocarbon Processing* 88 (1): 37–42. https://www.proquest.com/trade-journals/what-is-outlook-building-new-lng-facilities/docview/225205938/se-2?accountid=1038.

Priemus, H., B. Flyvbjerg, and B. van Wee, eds. 2008. *Decision-Making on Mega-Projects Cost–Benefit Analysis, Planning, and Innovation*. Cheltenham: Edward Elgar Publishing Limited.

Priemus, H., and B. van Wee. 2013. *International Handbook on Mega-Projects*. Cheltenham: Edward Elgar Publishing Limited.

Rahmat, R.Y., M.M. Jais, M.H.M. Rahim, M.A. Azil, S. Velon, and N.H. Nawawi. 2022. "Utilizing LNG Stochastic Model for Better Cost Predictability and Faster Decision Making During Front End Loading Stage." Paper Presented at the ADIPEC, Abu Dhabi, UAE. https://doi.org/10.2118/210923-MS.

Ram, J., D. Corkindale, and M. Wu. 2013. "Implementation Critical Success Factors (CSFs) for ERP: Do they Contribute to Implementation Success and Post-Implementation Performance?" *International Journal or Production Economics* 1 (144): 157–174. https://doi.org/10.1016/j.ijpe.2013.01.032.

Rangnow, D. 1994. "Rethinking Supply Acquisition." *Natural Gas* 10 (11): 24–27. https://doi.org/10.1002/gas.3410101109.

Ray, S.C., S.C. Kumbhakar, and P. Dua. 2015. *Benchmarking for Performance Evaluation, A Production Frontier Approach*. India: Springer Science Business Media.

Raydugin, Y. 2012. "Quantifying Unknown Unknowns in an Oil and Gas Capital Project." *International Journal of Risk and Contingency Management* 1 (2): 29–42. http://dx.doi.org/10.4018/ijrcm.2012040103.

Raydugin, Y. 2013. *Project Risk Management: Essential Methods for Project Teams and Decision Makers*. Hoboken: John Wiley & Sons, Inc.

Remington, K., and J. Pollack. 2012. *Tools for Complex Projects*. Farnham: Gower Publishing Ltd.

Rolstadas, A., P.W. Hetland, G.F. Jergeas, and R.E. Westney. 2011. *Risk Navigation Strategies for Major Capital Projects: Beyond the Myth of Predictability*. London: Verlag Limited.

Sainati, T., N. Brookes, and G. Locatelli. 2017. "Special Purpose Entities in Megaprojects: Empty Boxes or Real Companies?" *Project Management Journal* 48 (2): 55–73. https://doi.org/10.1177/875697281704800205.

Sakmar, S.L. 2013. *Energy for the 21st Century: Opportunities and Challenges for Liquefied Natural Gas (LNG)*. Cheltenham: Edward Elgar Publishing Limited.

Sanderson, J. 2012. "Risk, Uncertainty and Governance in Megaprojects: A Critical Discussion of Alternative Explanations." *International Journal of Project Management* 30 (4): 432–443. https://doi.org/10.1016/j.ijproman.2011.11.002.

Schauer, F. 2003. *Profiles, Probabilities, and Stereotypes*. London: Harvard University Press.

Schonberger, R.J. 2008. "Lean Performance Management (Metrics don't Add Up)." *Cost Management* 22 (1): 5–10. https://www.proquest.com/magazines/lean-performance-management-metrics-dont-add-up/docview/209699654/se-2?accountid=10382.

Schwarz, N. 1995. "What Respondents Learn from Questionnaires: The Survey Interview and the Logic of Conversation." *International Statistical Review* 63 (2): 153–168. https://doi.org/10.2307/1403610.

Shenhar, A., and D. Dvir. 2007. "Project Management Research – The Challenge and Opportunity, 1st ed." *Project Management Journal* 2 (38): 93–99. Project Management Institute. https://doi.org/10.1177/875697280703800210.

Shingan, B., N. Banerjee, M. Kumawat, P. Mitra, V. Parthasarthy, and V. Singh. 2024. "Liquified Natural Gas Value Chain: A Comprehensive Review and Analysis." *Chemical Engineering Technology* 47 (3): 430–447. https://doi.org/10.1002/ceat.202300304.

Simmons, D.R., C. McCall, and N.A. Clegorne. 2020. "Leadership Competencies for Construction Professionals as Identified by Construction Industry Executives." *Journal of*

Construction Engineering and Management 146 (9): 1–10. https://doi.org/10.1061/(ASCE)CO.1943-7862.0001903.

Singh. B., S. Grover, and V. Singh. 2017. "Evolution of Benchmarking Attribute for Service Quality Using Multi Attitude Decision Making Approach." *International Journal of Systems Assurance Engineering and Management* 8 (Suppl. 2): 617–630. https://doi.org/10.1007/s13198-016-0485-4.

Smith, C.A., and A.M. Herscowitz, eds. 2016. *Understanding Natural Gas and LNG Options*. https://www.energy.gov/sites/prod/files/2016/12/f34/Understanding%20Natural%20Gas%20and%20LNG%20Options.pdf.

Sobieraj, J., and D. Metelski. 2021. "Quantifying Critical Success Factors (CSFs) in Management of Investment-Construction Projects: Insights from Bayesian Model Averaging." *Buildings* 11 (8): 1–29. https://doi.org/10.3390/buildings11080360.

Söderlund, J., S. Sankaran, and C. Biesenthal. 2017. "The Past and Present of Megaprojects." *Project Management Journal* 48 (6): 5–16. http://journals.sagepub.com/doi/pdf/10.1177/875697281704800602.

Stapenhurst, T. 2009. *The Benchmarking Book: A How-to-Guide to Best Practice for Managers and Practitioners*. Oxford: Butterworth-Heinemann.

Stingl, V., and J. Geraldi. 2017. "Errors, Lies and Misunderstandings: Systematic Review on Behavioural Decision Making in Projects." *International Journal of Project Management* 35 (2): 121–135. https://doi.org/10.1016/j.ijproman.2016.10.009.

Stoklosa, R.T. 1999. "Practical Application of Environmental Risk Management – Gorgon LNG Project Case Study." *APPEA Journal* 39 (1): 606–621. https://doi.org/10.1071/AJ98040.

Taleb, N.N. 2004. "The Black Swan: Why Don't We Learn that We Don't Learn?" Paper Presented at the Highland Forum 23, Las Vegas, USA, November 2003. https://citeseerx.ist.psu.edu/viewdoc/download?doi=10.1.1.418.760&rep=rep1&type=pdf.

Tanaka, H. March 2014. "Toward Project and Program Management Paradigm in the Space of Complexity: A Case Study of Mega and Complex Oil and Gas Development and Infrastructure Projects." *Procedia – Social and Behavioural Sciences* 119: 65–74. https://doi.org/10.1016/j.sbspro.2014.03.010.

Tonon, G. 2015. "Integration of Qualitative and Quantitative Methods in Quality of Life Studies." In Tonon, G., ed. *Qualitative Studies in Quality of Life. Social Indicators Research Series*, vol. 55. Cham: Springer. https://doi.org/10.1007/978-3-319-13779-7_4 https://link.springer.com/chapter/10.1007/978-3-319-13779-7_4.

Toor, S., and S.O. Ogunlana. 2010. "Beyond the 'Iron Triangle': Stakeholder Perception of Key Performance Indicators (KPIs) for Large-Scale Public Sector Development Projects." *International Journal of Project Management* 28 (3): 228–236. https://doi.org/10.1016/j.ijproman.2009.05.005.

Trimble, G., and E. Lester. 2014. *Project Management, Planning and Control*. 6th ed. UK: Elsevier.

Turner, R., and R. Muller. 2005. "The Project Manager's Leadership Style as a Success Factor on Projects: A Literature Review." *Project Management Journal* 36 (2): 49–61. https://doi.org/10.1177/875697280503600206.

Tusiani, M.D., and G. Shearer. 2007. *LNG: A Nontechnical Guide*. Tulsa: PennWell Corporation.

Twomey, D. 28 June 2012. "Deloitte says LNG Fuelling Aust economic boom." *Econews*. http://econews.com.au/17597/deloitte-says-lng-fuelling-aust-economic-boom.

Victor, D.G., A.M. Jaffe, and M.H. Hayes, eds. 2006. *Natural Gas and Geopolitics From 1970 to 2040*. Cambridge: Cambridge University Press.

Weems, P.R., and M. Hwang. 2013. "Overview of Issues Common to Structuring, Negotiating and Documenting LNG Projects." *The Journal of World Energy Law & Business* 6 (4): 267–299. https://doi.org/10.1093/jwelb/jwt016.

Westland, J. 2007. *The Project Management Life Cycle; A Complete Step-by-Step Methodology for Initiating, Planning, Executing & Closing a Project Successfully*. London: Kogan Page Limited.

Westney, R. 2008. "The Westney Advisor." In *Are Stages and Gates Destroying Predictability? The Unintended Consequences of Front-End Loading.* Houston: Westney Consulting Group, Inc.

Williams, T. 2002. *Modelling Complex Projects*. Chichester: John Wiley & Sons.

Winch, G.M. 2010. *Managing Construction Projects: An Information Processing Approach*. 2nd ed. Oxford: Wiley-Blackwell.

Wuni, I.Y., and G. Shen. 2020. "Barriers to the Adoption of Modular Integrated Construction: Systematic Review and Meta-Analysis, Integrated Conceptual Framework, and Strategies." *Journal of Cleaner Production* 249. Science Direct: Elsevier. https://doi.org/10.1016/j.jclepro.2019.119347.

Yun, S., J. Choi, D. Oliveira, and S. Mulva. 2016. "Development of Performance Metrics for Phase-Based Capital Project Benchmarking." *International Journal of Project Management* 34 (3): 389–402. https://doi.org/10.1016/j.ijproman.2015.12.004.

Zairi, M. 2011. *Effective Management of Benchmarking Projects: Practical Guidelines & Examples of Best Practice*. Oxford: Butterworth-Heinemann.

Appendixes

Appendix 1 – Multifaceted Mega/Giga Projects Roadmap

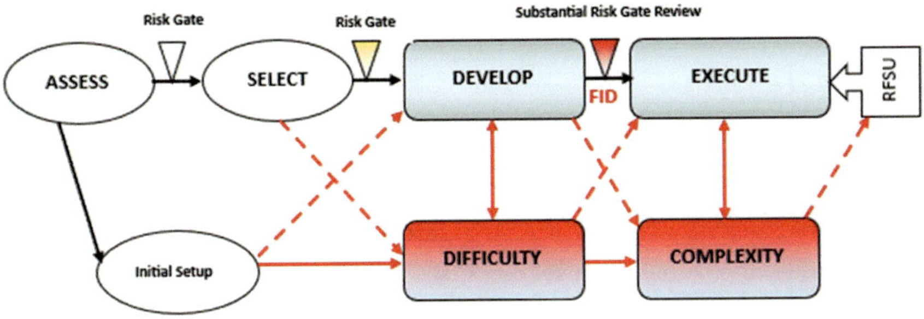

Complexity of Mega/Giga Projects

Appendix 2 – Overall Roadmap to RFSU

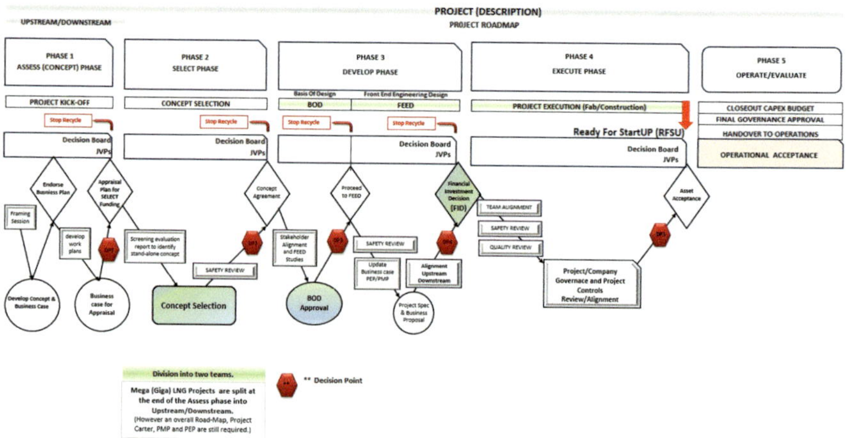

Index

Note: Page numbers in **bold** refer to tables; page numbers in *italics* refer to figures.